Introduction to Modern Navigation Systems

Esmat Bekir

World Scientific

NEW JERSEY • LONDON • SINGAPORE • BEIJING • SHANGHAI • HONG KONG • TAIPEI • CHENNAI

Published by

World Scientific Publishing Co. Pte. Ltd.
5 Toh Tuck Link, Singapore 596224
USA office: 27 Warren Street, Suite 401-402, Hackensack, NJ 07601
UK office: 57 Shelton Street, Covent Garden, London WC2H 9HE

British Library Cataloguing-in-Publication Data
A catalogue record for this book is available from the British Library.

INTRODUCTION TO MODERN NAVIGATION SYSTEMS

ISBN-13 978-981-270-765-9
ISBN-10 981-270-765-4
ISBN-13 978-981-270-766-6 (pbk)
ISBN-10 981-270-766-2 (pbk)

Printed in Singapore.

To my Lord

with humility and love

Preface

The goal of this text is to concisely present the mathematical blocks needed for implementing the main body of a strapped down Inertial Navigation System (INS) in a manner that provides a mental image of the contribution of each block and their interrelation. The text describes the makeup of each block and provides the derivation of its equations. Towards this objective, when the need for clarifying or justifying a certain idea arises, it is presented in an appendix so as not to interfere with the flow of the main ideas.

This treatment should benefit both the novice as well as a practitioner in the field. For a journeyman in the area of navigation, this book can be used to pinpoint the equations that are the basis of such a system, how they are developed and how they are implemented. Those with more experience may use this book as a quick reference guide.

What is navigation anyway? It is the ability to set the course of a ship to move between two desired locations. To do that the navigator must be able to know his location and set the velocity vector towards the desired destination. Thus the prime function of a navigation system is determining the craft's position and velocity.

We will be primarily concerned herein with a special type of navigation: inertial navigation. And why inertial navigation in particular? Inertial systems are self-contained: they are independent of weather conditions and are operable anywhere in seas, underwater, lands, tunnels, or in air. Short of a reliable source of power, they can work almost indefinitely.

If the Earth were flat, inertial navigation algorithms would have been a lot easier. Because navigation usually is on or close to Earth, a spherical body that rotates about itself, we will soon find ourselves entangled in discussing two different elements at the same time: developing the mathematical algorithms and describing the pertinent physics of the Earth.

We have devoted the first three chapters to introduce the mathematical foundation for developing the algorithms. These algorithms rely heavily on vector and matrix notations, and for that, vector and matrix properties are introduced in Chapter 1. On developing the equations, we will discover that our variables of interest are represented in different coordinate systems. Obviously this creates the need of moving from one coordinate system to another and thus this concept is discussed in Chapter 2. For further clarity, Chapter 3 introduces the most common approaches used in performing coordinate transformations. There we discuss these approaches and their relationships amongst one another.

We discuss the physical properties of the Earth in Chapter 4. At this point, armed with the mathematical tools and the geometrical properties of Earth, we develop the inertial navigation equations from first principles in Chapter 5. This yields a set of continuous time differential equations that should be solved to yield the navigation solution. Mechanizing and implementing these equations on a digital computer is introduced in Chapter 6. We discover one of the drawbacks of inertial navigation systems: unreliable vertical channel. This means that they cannot be relied on to provide altitude or vertical velocity.

Typically, the navigation system for aircraft is usually complemented with some sort of altimeter. Integrating the altimeter measurements with the INS is discussed therein. Our reliance on air data for aiding the navigation system does not end at this point: the INS cannot estimate wind speed. Chapter 7 is devoted to the discussion of air data computations and their use for computing rate of climb/decent and relative airspeed.

Using the legendary lines of longitude and latitude to locate a craft location on the surface of the Earth introduces a peculiar navigational phenomenon: all longitude lines meet at the two polar points of the Earth. Solving navigation equations near these two points can prove to be mathematically cumbersome. Despite the rarity of these events, considerable attention has been given to avert the consequences of such an occurrence. The wander azimuth angle is one such classical technique and it is critiqued in Chapter 8. An innovative simple algorithm for navigating in the polar circle is introduced shortly thereafter.

Two problems remain to be solved. The first addresses the alignment problem. Simply stated, it is determining the initial conditions of the differential equations that were developed in Chapter 5. The second deals with the real life factors: no matter how expensive the inertial sensors are they have errors that must be estimated. Solutions to these problems depends on the inertial sensor level accuracy onboard the specific craft.

To elaborate, we may have noticed that implementing the INS equations not only provide the craft location and speed, but also its attitude and heading. Some applications use navigation grade sensors to estimate all the above parameters. But others utilize low-grade inexpensive sensors focused on estimating only the attitude and heading. In so doing, these applications – called Attitude and Heading Reference Systems (AHRS) – forgo estimating the location and velocity.

Chapter 9 addresses the alignment problem for navigation systems. In Chapter 10, we discuss the AHRS systems, introduce their pertinent alignment algorithms, and estimate sensor errors. Often, an AHRS complements its inertial system with magnetic detectors and utilizes mathematical algorithms to estimate the relatively large errors introduced by the inertial systems.

A system that utilizes navigation grade sensors could enhance its performance by using aids such as a Global Positioning System (GPS). Like the INS, the GPS provides position and velocity. These two, INS and GPS, when mathematically fused with a Kalman filter, can be used to estimate the inaccuracies due to the sensors of the former system. These equations are developed in Chapter 11.

This book could not be a reality without the help and support from many friends and colleagues. I am very grateful to A. Bekir and H. Aleem for reviewing the script and suggesting many improvements. I would like to thank Dr. M. Hafez, Dr. M.A.M. Ali for support and encouragements. Special thanks to Ms. Kim Tan for her guidance throughout the publishing process. Last but not least I am very grateful to my family for their love and support during the entire time for writing the book.

Contents

Preface		vii
Introduction		1
1.	Vectors and Matrices	7
1.1	Introduction	7
1.2	Vector Inner Product	9
1.3	Vector Cross Products and Skew Symmetric Matrix Algebra	10
2.	Coordinate Transformation between Orthonormal Frames	17
2.1	Introduction	17
2.2	Direction Cosine Matrices	18
2.3	The Direction Cosine Matrix is a Unitary Matrix	20
2.4	The Direction Cosine Matrix is a Transformation Matrix	21
2.5	DCM Fixed Axis	24
2.6	The Rotation Matrix	26
2.7	Inner and Outer Transformation Matrices	29
2.8	The Quaternion	32
3.	Forms of the Transformation Matrix	35
3.1	Introduction	35
3.2	Simple Frame Rotations	36
3.3	Euler Angles	37
3.4	Rotation Vector	38
3.5	Quaternion	39
3.6	Simple Quaternions	43
3.7	Conversion between Forms	45
3.7.1	Conversion between DCM and Euler	45
3.7.2	Conversion between DCM and Quaternion	45
3.7.3	Conversion between Euler Angles and Quaternion	47

3.8 Dynamics of the Transformation Matrix 47
3.8.1 DCM Differential Equation 48
3.8.2 Quaternion Differential Equation 50
3.8.3 Rotation Vector Differential Equation 52
3.8.4 Euler Angles Differential Equation 55

4. Earth and Navigation 58
4.1 Introduction 58
4.2 Earth, Geoid and Ellipsoid 59
4.3 Radii of Curvature 63
4.4 Earth, Inertial and Navigation Frames 65
4.5 Earth Rate 67
4.6 The Craft Rate $\boldsymbol{\omega}_{en}^{n}$ 67
4.7 Solution of the DCM $\mathbf{C}_{e}^{n}$ 70
4.8 Gravitational and Gravity Fields 70

5. The Inertial Navigation System Equations 75
5.1 Introduction 75
5.2 Body Frame of Reference 76
5.3 Inertial Sensors 77
5.3.1 The Accelerometer 77
5.3.2 The Rate Gyro 78
5.4 The Attitude Equation 78
5.5 The Navigation Equation 80
5.6 Navigation Equations Computational Flow Diagram 83
5.7 The Navigation Equation in Earth Frame 84

6. Implementation 86
6.1 Introduction 86
6.2 The Rotation Vector Differential Equation 87
6.3 The Attitude Equation 92
6.4 The Craft Velocity Equation 95
6.5 The Craft Position Equation 99
6.6 The Vertical Channel 101

7. Air Data Computer 104
7.1 Introduction 104
7.2 US Standard Atmosphere 1976 105
7.3 Pressure Altitude 107
7.4 Vertical Channel Parameter Estimation Using Inertial and Air Data 111
7.5 Density Altitude 116

7.6 Altitude (Descend /Climb) Rate 117
7.7 Air Speed 117
7.8 Indicated Air Speed (IAS) 119

8. Polar Navigation 121
8.1 Introduction 121
8.2 The Wander Azimuth Navigation 123
8.3 Prospective of the Wander Azimuth Approach 126
8.4 Polar Circle Navigation Algorithm 128
8.5 Alternative Polar Circle Navigation Frame 132

9. Alignment 136
9.1 Introduction 136
9.2 IMU Alignment 137
9.3 Alternative Algorithm for $\mathbf{C}_n^b$ 144
9.4 Estimation of the Accelerometer and Gyro Biases 149
9.5 Effects of Biases on Estimate of $\mathbf{C}_n^b$ 150

10. Attitude and Heading Reference System 152
10.1 Introduction 152
10.2 Attitude Initialization 152
10.3 Heading Initialization 155
10.4 Gyro Drift Compensation 159
10.5 G Slaving 160
10.5.1 X-Gyro Bias 160
10.5.2 Y-Gyro Bias 162
10.5.3 Z-Gyro Bias 163
10.6 Alternative Approach for Gyro Drift Compensation 163
10.7 Maneuver Detector 165
10.7.1 Rate Gyro Threshold Selection 165

11. GPS Aided Inertial System 167
11.1 Introduction 167
11.2 Navigation Frame Error Equation 168
11.2.1 Craft Rate Error $\delta\omega_{en}^n$ 169
11.2.2 Earth Rate Error $\delta\omega_{ie}^n$ 170
11.2.3 Position Errors 171
11.2.4 Attitude Error 173
11.2.5 Gravity Error 176
11.2.6 Velocity Error 177
11.2.7 Navigation Frame Error State Equation 179
11.2.8 Error Block Diagram 179

11.3 Earth Frame Error Equations 180
11.3.1 Attitude Error 181
11.3.2 Velocity Error 182
11.3.3 Position Error 183
11.3.4 Earth Frame Error State Equation 183
11.4 Inertial Sensors Error Models 183
11.5 The Global Positioning System 187
11.6 Mechanization of the INS/GPS Equations 191

Appendix A. The Vector Dot and Cross Products 194

Appendix B. Introduction to Quaternion Algebra 197

Appendix C. Simulink® Models 202

Appendix D. Ellipse Geometry 206

Appendix E. Vector Dynamics 213

Appendix F. Derivation of Air Speed Equations 219

Appendix G. DCM Error Algebra 222

Appendix H. Kalman Filtering 226

Index 237

Introduction

Where am I? Sound familiar? We travel to some destination then we realize we lost our way. We agonize upon it when we overshoot or undershoot our destination. Or when we take the wrong route or the wrong freeway exit and find ourselves in a foreign territory. Quite often these incidents end up without incident, but in few occurrences they could result in unhappy endings. This is not a modern day problem but is something that man has lived with since the beginning of time. People needed to migrate to different lands in search of food and water. They needed to travel to trade with other people living in different areas. In so doing they did not migrate or travel en masse; they probably sent scouts to explore the unknown territories who had to return to inform their tribes with their findings. The scouts must have used some landmarks or followed some terrains so they could return safely. But what would they have done when the land they traveled over had no distinguishing features. They must have established some sort of bearing that could guide them in their back and forth trips. Similar measures must have been used when sailing the open seas; a ship with an unknown bearing was a vessel of the doomed.

The priority of navigation with the ancients has not changed over several millenniums and is just as relevant as it is with today's field of modern navigation. It is centered upon one core, unshakeable theme: location, location, location. If you want to move from one place to another, the first thing you need to know is your location. Next you should be able to set the velocity vector of your transportation means towards the desired destination.

Navigation is derived of the Latin verb *navigare*, to sail, which also is derived of '*navis*' a ship. Navigation has been the science, or very much the art, of determining the position and the velocity of a craft, whether a car, a ship or an airplane. It is of little wander that navigation become intertwined with sea and ships because it, by large means, evolved onboard of ships facing the enormous challenges of charting their courses. There are certain spots in water that are too shallow, rocky or turbulent that an experienced sailor would like to avert. Early travelers identified their locations and charted their routs by landmarks. When they got into water they, most probably, hugged very closely to the shoreline.

Navigation rose to higher level of sophistication when ancient Greeks realized that the Earth is spherical and when Eratosthenes of Alexandria measured its radius. They drew geographical maps superimposed on latitude and longitude quadrants.

Early on, people were aware that the sun and the stars can guide them when traveling long distances in the desert or in the sea. They knew that the altitude of Polaris, (the north star), indicates the latitude at which they locate. Near the equator (when Polaris is on the horizon and is difficult to see) or in daytime they had other means. They realized that at any point on Earth the sun attains its highest altitude at noon and this is when it points to the north and with a simple adjustment they can relate this altitude to the latitude at that point. Therefore, day or night, they could figure out the latitude. Navigators devised simple means for charting their routes: determine the initial latitude at the location they embark, sail north or south to the latitude of the desired destination and finally move east or west till they encounter their destination.

The latitude approach for navigation came into place because means for measuring the latitude were possible. Initially they came in the shape of the astrolabe, a simple apparatus, for measuring the altitude of the stars. The astrolabe depended on measuring the star altitude with respect to the horizon, but this meant poor measurements if the boat is rocking for bad weather or any reason. The astrolabe has evolved into the sextant, a more accurate tool [1].

Determining the latitude has solved one parameter of the navigation equation and it remained to determine the longitude. The magnetic needle is discovered, but probably it was not of much help because it would not point to the true geographic north, nor did it maintain a constant error. The magnetic compass would be of great use when magnetic variations, the difference between magnetic north and the true north, were tabulated.

Crude means for measuring the ship speed became also available that come in the form of chip log [2]. The means took the shape of a relatively heavy slug is attached to a long rope of about 700 feet length, that was knotted every 47 feet and 3 inches. If the sailor wanted to measure the boat speed, he would toss the slug into the water and let the rope slip in his hands. He would wait for a 28 seconds during which he is counting the number of knots that slips between his fingers. The time was measured by nothing more than the legendary sandglass. Interestingly, this is why the ship speed is still currently measured in 'knots'.

Between the sextant, the magnetic compass and the ship speed, navigators were able to chart their ship courses to move to different places. It should be remembered that these computations were not accurate and possibly crude. A rolling ship would certainly prevent collecting accurate sextant measurements as the sextant must be steady and resting on a perfectly horizontal surface. Crude, may be, but good enough to go to places. With these tools Christopher Columbus set sail to the Americas and charted his way back home.

Columbus trip in the final years of the 15th century was a monumental epoch in navigation; it opened a new era for longer trips that lasted longer periods in uncharted waters to uncharted lands. Ships tended to become larger to carry more people and larger cargos to sail between the new lands and Europe. Older methods and navigation tools resulted in larger tragedies and needed to be improved.

It was not until the 18th century that navigation got a great boost when John Harrison invented and developed the 'chronometer' and perfected it to measure time with an error of less than .1 seconds per day. Recall earlier that sun at noon points to the north and that, in a way, points to local longitude. Suppose a navigator set sail from a point on the Greenwich line (i.e. 0 longitude) and purposely sets the clock at noon to 12:00PM. Few days later, the clock is found to read 11:00 at noontime, the navigator would immediately know that the local longitude is 15 degrees east. This is because the Earth is girded with 360 longitude lines and since the Earth makes a complete circle every 24 hours, noon will differ by one hour every (360/24) 15 degrees of longitude. With Harrison's chronometer, British naval officer and explorer James Cooks was able in 1772-1775 to circumnavigate the Earth [3]. From information he gathered on his voyage, Cook completed many detailed charts of the world that completely changed the nature of navigation.

More accurate and much lighter sextants continued to evolve. However rough waters, which causes the ship to roll unsteadily, and fog, which causes the true horizon to be obscured, rendered the sextants very difficult to use under these conditions. Tools were needed to steady the sextant. Nothing but gyroscopic action that could do the job. Indeed it was a very fortunate moment when the roads of navigation and gyroscopes crossed and opened a new page in the history of navigation.

A gyroscope, basically, is a top that when rotates at very large rates, its upper surface would be horizontal and its axis of rotation and the Earth's axis make a plane that point to the north. What's more is that if the surface, on which the top spins, is disturbed the top will restore its phenomenon unless it is disturbed once more [4]. Amazing isn't it! It worth noting to mention that it was Foucault that named it 'gyroscope' in 1852. Not much time later when Admiral Fleuriais developed the gyroscopic sextant in 1885. But this is when the sextant reached its great year and started to give way to the gyro. Few factors has hastened the development of the new gyro: Larger ships made with iron alloys rendered the use of magnetic compassed very difficult, and temporal navigation and orientation tools – magnetic compasses and sextants – in submersibles were of little use.

The early days of the 20th century witnessed the dawn of the airborne aviation, and with it came the need for navigation tools on board of these flying machines. The legacy compass was still used to point direction. The new gyros, at the same time were being developed into the vertical and directional instruments to indicate the vertical and azimuth directions respectively. It turned out that the human sensing for the vertical direction while flying is very poor and hence the need for the vertical gyro to fly straight and level.

It all started by inventing and developing the 'Gyrocompass' – the practical instrument that indicates the direction to the north [5]. It performs well up to 80 degrees latitude. The gyrocompass was invented and developed by Dr. Hermann Anschuetz-Kaempfe in Kiel, Germany in 1908 [6]. "Compared with the conventional marine-type magnetic compass, the gyrocompass was far more accurate, being capable of indicating the north within a small fraction of one degree. It has no variation error" [7]. The gyrocompass was also developed in the USA by Elmer Sperry whose company has produced many versions. These gyrocompasses were installed on board commercial and navy ships.

The era of inertial navigation has started when gyros and accelerometers were used as a guidance tools in the German V2 rockets. With the end of the Second World War came a speedy activity for developing an inertial navigation system. This system comprises one triad of accelerometers and another triad of gyros, both are mounted on top of a platform called the stable element. The stable element is mounted on two or more orthogonal gimbals that allow it to have a complete three degrees of freedom orientation. With the gyros aid, the system is so designed that, the stable element will maintain its attitude with respect to the surface of the Earth. This attitude, as an example, is parallel to the local surface of the Earth and pointing to the north. With this arrangement, the accelerometer outputs are integrated once to give the craft velocity and integrated once more to give its location. Indeed it is a remarkable tool for navigation. Short of a reliable source of power, it can work almost indefinitely. With one of these systems the US Navy Nautilus navigated under ice to reach the North Pole in 1958.

This inertial navigation system is self-contained: it is independent of weather conditions and is operable anywhere – in seas, underwater, lands, tunnels, or in air. On the other hand it is very costly (in the order of $100,000 in the 1970s), bulky and comprises many delicate components, sensitive to temperature variations and hence is very expensive to maintain.

In 1956, W. Newell patented the idea of the Strapdown Inertial Navigation System (INS). The patent described the implementation of an INS strapped down of its gimbals and literally fixing the inertial platform to the body of the craft. The thought is to trade the mechanical orientation with an analytical one with use of onboard computer. True that inertial platform will not maintain its attitude with respect to the surface of Earth and as such the accelerations measured by the accelerometers can't be integrated to obtain the velocity and position of the vehicle. But on the other hand the attitude of the craft can be tracked accurately by the mounting gyros. This demanded the use of highly accurate gyros and the computer that can perform these intensive computations. At the time of the patent, lack of digital computers of reasonable size delayed the development of such system till the early 1970's. When they became available, a new page of modern navigation has begun.

It is the subject of this book to introduce the mathematical and physical concepts of the modern inertial navigation system. Also to formulate the equations that should be implemented on the computer to deduce the navigation solution.

References

1. The History of the Sextant: http://www.mat.uc.pt/%7Ehelios/Mestre/Novemb00/H61iflan.htm
2. chiplog.htm Backstaff Instruments: Period Navigation Inst., Marblehead, MA
3. http://www.users.tpg.com.au/vmrg/James%20Cook.html
4. H. W. Sorg, 'From Serson to Draper – Two Centuries of Gyroscopic Development', Journal of the Institute of Navigation, Vol 23, No 4, pp 313-324.
5. American Practical Navigation: An Epitome of Navigation, Defense Mapping Agency Hydrographic/Topographic Center, Vol I, 1984.
6. http://www.raytheonmarine.de/highseas/company_information/profile_english.html
7. Paul Savet Editor, Gyroscopes: Theory and Design with Applications to Instrumentation, Guidance and Control, McGraw Hill, NY, New York, 1961.

Chapter 1

Vectors and Matrices

1.1 Introduction

Vectors and matrices are used extensively throughout this text. Both are essential as one cannot derive and analyze laws of physics and physical measurements without vectors and one cannot process these measurements in a digital computer without matrices. Further, each has powerful features that when combined can result in considerable derivation simplifications. We shall highlight the main similarities and the subtle differences between them.

In this context we will be concerned with Euclidean vector spaces for which the inner product is defined. In n-dimensional Euclidean vector spaces it is possible to construct a set of n orthogonal unit vectors $\mathbf{r}_1, \mathbf{r}_2, \ldots, \mathbf{r}_n$. As such, an arbitrary vector $\mathbf{v}$ in this space is represented by

$$\mathbf{v} = v_1\mathbf{r}_1 + v_2\mathbf{r}_2 + \cdots + v_n\mathbf{r}_n \tag{1.1}$$

where $v_1, v_2, \ldots, v_n$ are the scalar coordinates of $\mathbf{v}$.

In the special case of three dimensional vector space, the unit vectors $\mathbf{r}_1, \mathbf{r}_2, \mathbf{r}_3$ can be graphically represented by a set of three orthogonal axes. Hence, v_1, v_2, v_3 (the coordinates of the 3-dimensional vector $\mathbf{v}$) are the projections of $\mathbf{v}$ along $\{\mathbf{r}_1, \mathbf{r}_2, \mathbf{r}_3\}$.

A matrix on the other hand, is an array of n rows and m columns, of $n \times m$ numbers [1,2]. For example, a 3×3 matrix $\mathbf{A}$ is represented by

$$\mathbf{A} = \begin{bmatrix} a_{11} & a_{12} & a_{13} \\ a_{21} & a_{22} & a_{23} \\ a_{31} & a_{32} & a_{33} \end{bmatrix}$$

The transpose of a $i \times j$ matrix **B** is a $j \times i$ matrix, denoted by **B**′, in which the rows and columns of **B** trade places. For example, the transpose of the above matrix **A** is given by

$$\mathbf{A}' = \begin{bmatrix} a_{11} & a_{21} & a_{31} \\ a_{12} & a_{22} & a_{32} \\ a_{13} & a_{23} & a_{33} \end{bmatrix}$$

In particular, a $n \times 1$ array is a column matrix and a $1 \times n$ array is a row matrix. Transposing a row matrix makes it a column matrix and vice versa.

To adapt vectors to matrix manipulations it is common to use row or column matrices whose components are the vector coordinates. When there is no confusion, column matrices will be used to denote vectors. So vector **v**, in Eq. (1.1), can be represented in matrix form by

$$\mathbf{v} = \begin{bmatrix} v_1 \\ v_2 \\ \vdots \\ v_n \end{bmatrix} \tag{1.2}$$

Implicit in the above equation that the components $v_1, v_2, \ldots, v_n$ are the coordinates of **v** along the vectors $\mathbf{r}_1, \mathbf{r}_2, \ldots, \mathbf{r}_n$.

1.2 Vector Inner Product

The inner product of the two real vectors **u** and **v** is defined by [3,4]

$$\mathbf{u}\cdot\mathbf{v} = u_1 v_1 + u_2 v_2 + \cdots + u_n v_n \tag{1.3}$$

where $\{u_1, u_2, \ldots, u_n\}$ and $\{v_1, v_2, \ldots, v_n\}$ are the coordinates of **u** and **v**. The length of vector **u**, also called its norm, is defined by the inner product

$$\mathrm{norm}(\mathbf{u}) = |\mathbf{u}| = \sqrt{\mathbf{u}\cdot\mathbf{u}} = \sqrt{u_1^2 + u_2^2 + \cdots + u_n^2} \tag{1.4}$$

We call **u** and **v** orthogonal if their inner product equals zero; if, in addition, each is of unit length then they are orthonormal. In particular the set of vectors $\mathbf{r}_1, \mathbf{r}_2, \ldots, \mathbf{r}_n$, introduced above, are orthonormal because

$$\mathbf{r}_i\cdot\mathbf{r}_i = 1 \quad \text{and} \quad \mathbf{r}_i\cdot\mathbf{r}_j = 0, \quad i, j = 1, 2, \ldots, n;\ i \neq j \tag{1.5}$$

Thus it can be seen from Eqs. (1.1) and (1.5) that the coordinates of vector **v** are given by

$$v_i = \mathbf{v}\cdot\mathbf{r}_i, \quad i = 1, 2, \ldots, n \tag{1.6}$$

In a three dimensional space the inner product has a physical significance. The cosine of the angle between vectors **u** and **v**, denoted by cos(**u, v**) is determined by the cosine law (see Appendix A)

$$\cos(\mathbf{u},\mathbf{v}) = \frac{\mathbf{u}\cdot\mathbf{v}}{|\mathbf{u}||\mathbf{v}|} \tag{1.7}$$

If **v** is a unit vector we get from Eqs. (1.6) and (1.7)

$$v_i = \mathbf{v} \cdot \mathbf{r}_i = \cos(\mathbf{v}, \mathbf{r}_i), \quad i = 1, 2, \ldots, n \tag{1.8}$$

When a vector set $\{\mathbf{r}_1, \mathbf{r}_2, \ldots, \mathbf{r}_n\}$ is represented in column matrices, then they are orthonormal if

$$\mathbf{r}_i' \mathbf{r}_i = 1 \quad \text{and} \quad \mathbf{r}_i' \mathbf{r}_j = 0, \quad i, j = 1, 2, \ldots, n; \; i \neq j \tag{1.9}$$

The Hermitian inner product of the complex vectors **u** and **v** is defined by

$$\mathbf{u} \cdot \mathbf{v}^* = \mathbf{u}' \mathbf{v}^* = u_1 v_1^* + u_2 v_2^* + \ldots + u_n v_n^* \tag{1.10}$$

where the prime denotes the matrix transpose and the * denotes the complex conjugate. A matrix is called square if its number of rows is the same as its number of columns. The identity matrix is a real square whose diagonal components are all ones and the rest are all zeros. For example the 3×3 identity matrix is

$$\mathbf{I} = \begin{bmatrix} 1 & 0 & 0 \\ 0 & 1 & 0 \\ 0 & 0 & 1 \end{bmatrix}$$

1.3 Vector Cross Products and Skew Symmetric Matrix Algebra

Suppose that $\{\mathbf{r}_1, \mathbf{r}_2, \mathbf{r}_3\}$ is an orthonormal vector set for a 3-dimensional vector space in which the two vectors **u** and **v** are given by

$$\mathbf{u} = u_1\mathbf{r}_1 + u_2\mathbf{r}_2 + u_3\mathbf{r}_3$$
$$\mathbf{v} = v_1\mathbf{r}_1 + v_2\mathbf{r}_2 + v_3\mathbf{r}_3$$

then the vector cross product of **u** and **v**, denoted by $\mathbf{w} = \mathbf{u} \times \mathbf{v}$, is defined by [5,6]

$$\mathbf{w} = (u_2v_3 - u_3v_2)\mathbf{r}_1 + (u_3v_1 - u_1v_3)\mathbf{r}_2 + (u_1v_2 - u_2v_1)\mathbf{r}_3 \quad (1.11)$$

Notice that the cross product of the two vectors **u** and **v** is another vector that is orthogonal to both **u** and **v**. In matrix notation **w** is given by

$$\mathbf{w} = \mathbf{u} \times \mathbf{v} = \begin{bmatrix} u_2v_3 - u_3v_2 \\ u_3v_1 - u_1v_3 \\ u_1v_2 - u_2v_1 \end{bmatrix}$$

It is shown in Appendix A that if **u** and **v** are unit vectors then the magnitude of their vector cross product, **w,** is the sine of the angle between them. Therefore if **k** is the unit vector along **w** then

$$\mathbf{w} = \mathbf{u} \times \mathbf{v} = \sin(\mathbf{u}, \mathbf{v})\mathbf{k} \quad (1.12)$$

A skew symmetric matrix **B** is a matrix with the property of **B′** = -**B**. A 3-dimensional skew symmetric matrix emulates the cross product operation and enables it to be expressed in matrix notation. The vector

$$\mathbf{b} = \begin{bmatrix} b_1 \\ b_2 \\ b_3 \end{bmatrix}$$

corresponds to the skew symmetric matrix defined by

$$\mathrm{S}(\mathbf{b}) = \tilde{\mathbf{b}} = \begin{bmatrix} 0 & -b_3 & b_2 \\ b_3 & 0 & -b_1 \\ -b_2 & b_1 & 0 \end{bmatrix} \tag{1.13}$$

The operator S and the tilde notation are identical and will be used interchangeably to denote skew symmetric matrices, even though the latter will be used whenever possible.

Properties of the Skew Symmetric Matrix

In the following it is assumed that **a**, **b** and **w** are three dimensional arbitrary vectors (column matrix) and α and β are arbitrary scalars.

1. Correspondence to vector cross product:

$$\tilde{\mathbf{b}}\mathbf{w} = \begin{bmatrix} 0 & -b_3 & b_2 \\ b_3 & 0 & -b_1 \\ -b_2 & b_1 & 0 \end{bmatrix} \begin{bmatrix} w_1 \\ w_2 \\ w_3 \end{bmatrix} = \begin{bmatrix} b_2 w_3 - b_3 w_2 \\ b_3 w_1 - b_1 w_3 \\ b_1 w_2 - b_2 w_1 \end{bmatrix} = \mathbf{b} \times \mathbf{w} \tag{1.14}$$

Thus operating the matrix product of $\tilde{\mathbf{b}}$ and **w** corresponds to the vector cross product of the **b** and **w**.

2. Linearity:

$$\alpha \mathrm{S}(\mathbf{a}) + \beta \mathrm{S}(\mathbf{b}) = \alpha \begin{bmatrix} 0 & -a_3 & a_2 \\ a_3 & 0 & -a_1 \\ -a_2 & a_1 & 0 \end{bmatrix} + \beta \begin{bmatrix} 0 & -b_3 & b_2 \\ b_3 & 0 & -b_1 \\ -b_2 & b_1 & 0 \end{bmatrix} \Rightarrow$$

$$\alpha \mathrm{S}(\mathbf{a}) + \beta \mathrm{S}(\mathbf{b}) = \begin{bmatrix} 0 & -\alpha a_3 - \beta b_3 & \alpha a_2 + \beta b_2 \\ \alpha a_3 + \beta b_3 & 0 & -\alpha a_1 - \beta b_1 \\ -\alpha a_2 - \beta b_2 & \alpha a_1 + \beta b_1 & 0 \end{bmatrix} \Rightarrow$$

$$\alpha \mathrm{S}(\mathbf{a}) + \beta \mathrm{S}(\mathbf{b}) = \mathrm{S}(\alpha \mathbf{a} + \beta \mathbf{b}) \tag{1.15}$$

3. Skewness

$$\tilde{\mathbf{b}}' = \begin{bmatrix} 0 & -b_3 & b_2 \\ b_3 & 0 & -b_1 \\ -b_2 & b_1 & 0 \end{bmatrix}' = \begin{bmatrix} 0 & b_3 & -b_2 \\ -b_3 & 0 & b_1 \\ b_2 & -b_1 & 0 \end{bmatrix} = -\tilde{\mathbf{b}}$$

Thus

$$\tilde{\mathbf{b}}' = -\tilde{\mathbf{b}} \tag{1.16}$$

4. Operating on self vector

$$\tilde{\mathbf{b}}\mathbf{b} = \mathbf{b} \times \mathbf{b} = \mathbf{0} \tag{1.17}$$

Lemma 1.1

$$\mathrm{S}(\mathbf{b} \times \mathbf{w}) = \tilde{\mathbf{b}}\tilde{\mathbf{w}} - \tilde{\mathbf{w}}\tilde{\mathbf{b}} \tag{1.18}$$

Proof: Since

$$\begin{aligned} \tilde{\mathbf{b}}\tilde{\mathbf{w}} &= \begin{bmatrix} 0 & -b_3 & b_2 \\ b_3 & 0 & -b_1 \\ -b_2 & b_1 & 0 \end{bmatrix} \begin{bmatrix} 0 & -w_3 & w_2 \\ w_3 & 0 & -w_1 \\ -w_2 & w_1 & 0 \end{bmatrix} \\ &= \begin{bmatrix} -b_2 w_2 - b_3 w_3 & b_2 w_1 & b_3 w_1 \\ b_1 w_2 & -b_3 w_3 - b_1 w_1 & b_3 w_2 \\ b_1 w_3 & b_2 w_3 & -b_1 w_1 - b_2 w_2 \end{bmatrix} \end{aligned} \tag{1.19}$$

The RHS matrix can be arranged in the form of

$$\tilde{\mathbf{b}}\tilde{\mathbf{w}} = -(\mathbf{w}'\mathbf{b})\mathbf{I} + \mathbf{w}\mathbf{b}' \tag{1.20}$$

Likewise

$$\tilde{\mathbf{w}}\tilde{\mathbf{b}} = -(\mathbf{b}'\mathbf{w})\mathbf{I} + \mathbf{b}\mathbf{w}' \tag{1.21}$$

Subtracting Eq. (1.21) from Eq. (1.20) gives

$$\begin{aligned}
\mathbf{w}\mathbf{b}' - \mathbf{b}\mathbf{w}' &= \tilde{\mathbf{b}}\tilde{\mathbf{w}} - \tilde{\mathbf{w}}\tilde{\mathbf{b}} \\
&= \begin{bmatrix} 0 & b_2 w_1 - b_1 w_2 & b_3 w_1 - b_1 w_3 \\ b_1 w_2 - b_2 w_1 & 0 & b_3 w_2 - b_2 w_3 \\ b_1 w_3 - b_3 w_1 & b_2 w_3 - b_3 w_2 & 0 \end{bmatrix} \qquad (1.22) \\
&= \mathrm{S}\begin{bmatrix} b_2 w_3 - b_3 w_2 \\ b_3 w_1 - b_1 w_3 \\ b_1 w_2 - b_2 w_1 \end{bmatrix} = \mathrm{S}(\mathbf{b} \times \mathbf{w})
\end{aligned}$$

Before we explore further properties of the skew matrix we now introduce the orthonormal matrix. A matrix $\mathbf{R} = [\mathbf{r}_1 \quad \mathbf{r}_2 \quad \ldots \quad \mathbf{r}_n]$ is called orthonormal if its columns are mutually orthonomal, or equivalently satisfies Eq. (1.9). A three-dimensional orthonormal matrix

$$\mathbf{C} = [\mathbf{c}_1 \quad \mathbf{c}_2 \quad \mathbf{c}_3] \tag{1.23}$$

have these properties

$$\begin{aligned} \mathbf{c}_1 \times \mathbf{c}_2 &= \mathbf{c}_3, \\ \mathbf{c}_2 \times \mathbf{c}_3 &= \mathbf{c}_1, \\ \mathbf{c}_3 \times \mathbf{c}_1 &= \mathbf{c}_2 \end{aligned} \tag{1.24}$$

Lemma 1.2 If **C** is 3×3 orthonormal matrix, then

$$\mathbf{C}\,\mathrm{S}(\mathbf{b})\mathbf{C}' = \mathrm{S}(\mathbf{C}\mathbf{b}) \tag{1.25}$$

Proof: Since

$$\mathbf{C}\tilde{\mathbf{b}}\mathbf{C}' = \begin{bmatrix} \mathbf{c}_1 & \mathbf{c}_2 & \mathbf{c}_3 \end{bmatrix} \begin{bmatrix} 0 & -b_3 & b_2 \\ b_3 & 0 & -b_1 \\ -b_2 & b_1 & 0 \end{bmatrix} \mathbf{C}'$$

Multiplying **C** by $\tilde{\mathbf{b}}$ and expanding **C′** yields

$$\mathbf{C}\tilde{\mathbf{b}}\mathbf{C}' = \begin{bmatrix} b_3\mathbf{c}_2 - b_2\mathbf{c}_3 & b_1\mathbf{c}_3 - b_3\mathbf{c}_1 & b_2\mathbf{c}_1 - b_1\mathbf{c}_2 \end{bmatrix} \begin{bmatrix} \mathbf{c}_1' \\ \mathbf{c}_2' \\ \mathbf{c}_3' \end{bmatrix}$$

Multiplying the two right hand side matrices and collecting terms gives

$$\mathbf{C}\tilde{\mathbf{b}}\mathbf{C}' = b_1(\mathbf{c}_3\mathbf{c}_2' - \mathbf{c}_2\mathbf{c}_3') + b_2(\mathbf{c}_1\mathbf{c}_3' - \mathbf{c}_3\mathbf{c}_1') + b_3(\mathbf{c}_2\mathbf{c}_1' - \mathbf{c}_1\mathbf{c}_2')$$

Applying Eq. (1.22) to the above equation yields

$$\mathbf{C}\tilde{\mathbf{b}}\mathbf{C}' = b_1\,\mathrm{S}(\mathbf{c}_2 \times \mathbf{c}_3) + b_2\,\mathrm{S}(\mathbf{c}_3 \times \mathbf{c}_1) + b_3\,\mathrm{S}(\mathbf{c}_1 \times \mathbf{c}_2)$$

Substituting from Eq. (1.24) in the above equation gives

$$\mathbf{C}\tilde{\mathbf{b}}\mathbf{C}' = b_1 \mathrm{S}(\mathbf{c}_1) + b_2 \mathrm{S}(\mathbf{c}_2) + b_3 \mathrm{S}(\mathbf{c}_3)$$

By virtue of the linearity of the S operator it follows that

$$\mathbf{C}\tilde{\mathbf{b}}\mathbf{C}' = \mathrm{S}(b_1\mathbf{c}_1 + b_2\mathbf{c}_2 + b_3\mathbf{c}_3) = \mathrm{S}\left(\begin{bmatrix}\mathbf{c}_1 & \mathbf{c}_2 & \mathbf{c}_3\end{bmatrix}\begin{bmatrix} b_1 \\ b_2 \\ b_3 \end{bmatrix}\right) = \mathrm{S}(\mathbf{Cb})$$

In this chapter we explored the little differences between the vector and matrix notations. In general, a vector represents a direction in the three dimensional Euclidean space and a magnitude. A matrix is a set of elements that are arranged in a specific manner. A vector cross product is a special operation pertains to vectors but can be emulated in matrix notations using the skew symmetric matrix. One advantage with matrix products is their associativity, $(\mathbf{AB})\mathbf{C} = \mathbf{A}(\mathbf{BC})$ which is not the case for vector cross products, $(\mathbf{a} \times \mathbf{b}) \times \mathbf{c} \neq \mathbf{a} \times (\mathbf{b} \times \mathbf{c})$. Awareness of these distinctions will allow us to move from one notation to the other as desired. In the following chapter, the usefulness of these tools will be very vivid as they allow us to describe vector rotations (that are given in vector notations) in terms of transformation matrices.

References

1. D. E. Bourne, P. C. Kendall, Vector Analysis, Allyn and Bacon Inc., Boston, MA, 1967.
2. R. Bellman, Introduction to Matrix Analysis, McGraw Hill, New York, New York, 1970.
3. R. Larson, R. Hostetler, B. Edwards, Calculus with Analytic Geometry, D. C. Heath and Company, Lexington, Ma, 1994.
4. D. Varberg, E. Purcell, Calculus with Analytic Geometry, Prentice Hall, Englewood Cliffs, New Jersey, 1992.

Chapter 2

Coordinate Transformation between Orthonormal Frames

2.1 Introduction

Consider a coordinate system with a set of three orthogonal axes centered on the Earth's geometric center: the z-axis passes through the North Pole and the x and y axes are in the equatorial plane and where the x-axis is along the Greenwich meridian. Could you imagine this coordinate system being used to locate objects on Earth? Imagine an aircraft approaching the Los Angeles Airport and identifying its location relative to this coordinate system as (-2499km, 2952km, 3528km); can an airport traffic controller in the tower guide this aircraft to land safely? Yes, it can be done but it will be cumbersome and most importantly counterintuitive. It would be more useful and more intuitive if the aircraft had identified its location by its coordinates relative to a Cartesian coordinate system centered at the tower or by its longitude, latitude and altitude. To be sure, the Earth center coordinate system we have just described is very useful but for other purposes.

In general, there is no unique set of axes to define an n-dimensional space since any set of n orthonormal vectors can be used to represent arbitrary n-dimensional vectors. As will be seen, it is convenient to favor a specific frame to represent vectors over other frames. For example, it is a nightmarish process, except for trivial cases, solving the Newton's angular momentum equations using a fixed frame system. However, it is much easier to solve these equations using a frame system fixed to the body (where the angular moments of inertia would be constants).

Consequently, we should be able to transform a vector from one frame to another and hence it is important to determine the relationship between these different representations. For our purposes, discussion will be constrained to three-dimensional spaces and use of graphical representations as much as possible.

2.2 Direction Cosine Matrices

Suppose that $\boldsymbol{R} = \{\mathbf{r}_1, \mathbf{r}_2, \mathbf{r}_3\}$ and $\boldsymbol{S} = \{\mathbf{s}_1, \mathbf{s}_2, \mathbf{s}_3\}$ are two sets of orthonormal vectors, and $\boldsymbol{R}$ is our reference frame. Further, let l_1, m_1, and n_1 be the projections of vector $\mathbf{s}_1$ along $\{\mathbf{r}_1, \mathbf{r}_2, \mathbf{r}_3\}$ as depicted in Fig. (2.1). From Eq. (1.8) we see that

$$\begin{bmatrix} l_1 & m_1 & n_1 \end{bmatrix} = \begin{bmatrix} \cos(\mathbf{s}_1, \mathbf{r}_1) & \cos(\mathbf{s}_1, \mathbf{r}_2) & \cos(\mathbf{s}_1, \mathbf{r}_3) \end{bmatrix} \quad (2.1)$$

and thus the coordinates (l_1, m_1, n_1) are called the direction cosines of $\mathbf{s}_1$.

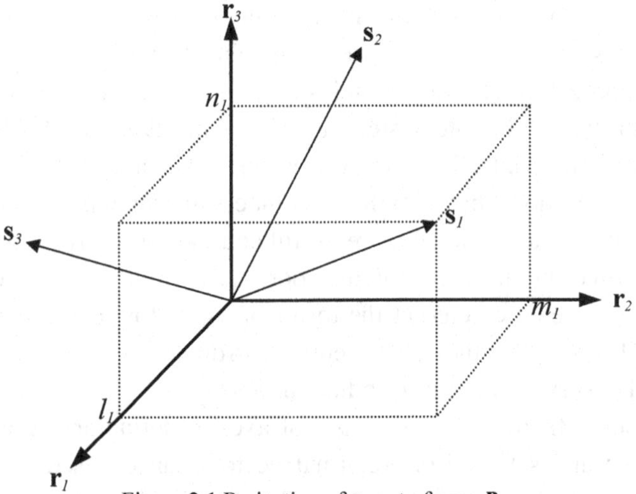

Figure 2.1 Projection of $\mathbf{s}_1$ onto frame $\boldsymbol{R}$

The vector $\mathbf{s}_1$ can be expressed in frame $\boldsymbol{R}$ as

$$\mathbf{s}_1 = l_1 \mathbf{r}_1 + m_1 \mathbf{r}_2 + n_1 \mathbf{r}_3 \quad (2.2)$$

Similar to $\mathbf{s}_1$, the projections of $\mathbf{s}_2$ and $\mathbf{s}_3$ are given by

$$\mathbf{s}_2 = l_2\mathbf{r}_1 + m_2\mathbf{r}_2 + n_2\mathbf{r}_3 \tag{2.3}$$

$$\mathbf{s}_3 = l_3\mathbf{r}_1 + m_3\mathbf{r}_2 + n_3\mathbf{r}_3 \tag{2.4}$$

The above three equations that represent the projection of $\{\mathbf{s}_1, \mathbf{s}_2, \mathbf{s}_3\}$ onto frame $\boldsymbol{R}$ can be combined in a matrix equation to get

$$[\mathbf{s}_1 \quad \mathbf{s}_2 \quad \mathbf{s}_3] = [\mathbf{r}_1 \quad \mathbf{r}_2 \quad \mathbf{r}_3]\begin{bmatrix} l_1 & l_2 & l_3 \\ m_1 & m_2 & m_3 \\ n_1 & n_2 & n_3 \end{bmatrix}$$

Transposing both sides gives

$$\begin{bmatrix} \mathbf{s}_1' \\ \mathbf{s}_2' \\ \mathbf{s}_3' \end{bmatrix} = \begin{bmatrix} l_1 & m_1 & n_1 \\ l_2 & m_2 & n_2 \\ l_3 & m_3 & n_3 \end{bmatrix}\begin{bmatrix} \mathbf{r}_1' \\ \mathbf{r}_2' \\ \mathbf{r}_3' \end{bmatrix} \tag{2.5}$$

Let $\mathbf{C}$ denote the array on the RHS of Eq. (2.5), then

$$\mathbf{C} = [\mathbf{c}_1 \quad \mathbf{c}_2 \quad \mathbf{c}_3] = \begin{bmatrix} l_1 & m_1 & n_1 \\ l_2 & m_2 & n_2 \\ l_3 & m_3 & n_3 \end{bmatrix} \tag{2.6}$$

Because each element in the above 3×3 array is a cosine term, it is called the direction cosine matrix (DCM). In the following Section we explore some intrinsic properties of a DCM.

2.3 The Direction Cosine Matrix is a Unitary Matrix

A unitary matrix is the matrix whose inverse is its transpose. From Eqs. (2.2)-(2.4) and using the orthonormality of the $\{\mathbf{r}_1, \mathbf{r}_2, \mathbf{r}_3\}$ set, we could show that

$$\mathbf{s}_i \cdot \mathbf{s}_i = l_i^2 + m_i^2 + n_i^2 = 1, \quad i = 1,2,3 \tag{2.7a}$$

$$\mathbf{s}_i \cdot \mathbf{s}_j = l_i l_j + m_i m_j + n_i n_j = 0, \quad i,j = 1,2,3;\, i \neq j \tag{2.7b}$$

Now the matrix product of Eq. (2.6) with its transpose is given by

$$\mathbf{CC}' = \begin{bmatrix} l_1 & m_1 & n_1 \\ l_2 & m_2 & n_2 \\ l_3 & m_3 & n_3 \end{bmatrix} \begin{bmatrix} l_1 & l_2 & l_3 \\ m_1 & m_2 & m_3 \\ n_1 & n_2 & n_3 \end{bmatrix}$$

$$= \begin{bmatrix} l_1^2 + m_1^2 + n_1^2 & l_1 l_2 + m_1 m_2 + n_1 n_2 & l_1 l_3 + m_1 m_3 + n_1 n_3 \\ l_2 l_1 + m_2 m_1 + n_2 n_1 & l_2^2 + m_2^2 + n_2^2 & l_2 l_3 + m_2 m_3 + n_2 n_3 \\ l_3 l_1 + m_3 m_1 + n_3 n_1 & l_3 l_2 + m_3 m_2 + n_3 n_2 & l_3^2 + m_3^2 + n_3^2 \end{bmatrix}$$

Substituting from Eq. (2.7) in the above gives

$$\mathbf{CC}' = \begin{bmatrix} 1 & 0 & 0 \\ 0 & 1 & 0 \\ 0 & 0 & 1 \end{bmatrix} = \mathbf{I} \tag{2.8}$$

which implies that $\mathbf{C}$ and $\mathbf{C}'$ are unitary matrices; each is the inverse of the other. Given, by definition, that a unitary matrix satisfies Eq. (2.8) implies that its columns (and rows) are an orthonormal set, which shows that a unitary matrix is also a DCM.

2.4 The Direction Cosine Matrix is a Transformation Matrix

A transformation matrix is a matrix that transforms the coordinates of a vector from one frame to another [1,2]. It will be shown that the DCM is the matrix that performs this transformation. Since the coordinates of a vector are dependent on the frame it is represented in, then an arbitrary vector **v** can be represented in frame ***R*** as

$$\mathbf{v} = (\mathbf{v}.\mathbf{r}_1)\,\mathbf{r}_1 + (\mathbf{v}.\mathbf{r}_2)\,\mathbf{r}_2 + (\mathbf{v}.\mathbf{r}_3)\,\mathbf{r}_3 \tag{2.9}$$

In matrix notation the above equation is expressed as

$$\mathbf{v} = \begin{bmatrix} \mathbf{r}_1 & \mathbf{r}_2 & \mathbf{r}_3 \end{bmatrix} \begin{bmatrix} v_{1r} \\ v_{2r} \\ v_{3r} \end{bmatrix} \tag{2.10a}$$

where

$$v_{ir} = \mathbf{v}.\mathbf{r}_i = \mathbf{r}_i^{\prime}\mathbf{v}, \qquad i = 1,2,3. \tag{2.10b}$$

To define the vector projection in the above equation, we used both the vector notation and the matrix notation (it will be needed later). Likewise in frame ***S***, the vector **v** is expressed as

$$\mathbf{v} = \begin{bmatrix} \mathbf{s}_1 & \mathbf{s}_2 & \mathbf{s}_3 \end{bmatrix} \begin{bmatrix} v_{1s} \\ v_{2s} \\ v_{3s} \end{bmatrix} \tag{2.11a}$$

where

$$v_{is} = \mathbf{v}.\mathbf{s}_i = \mathbf{s}_i'\mathbf{v}, \quad i = 1,2,3. \tag{2.11b}$$

To distinguish between the projections of **v** onto frames ***R*** and ***S*** their respective coordinates will be denoted as follows,

$$\mathbf{v}^r = \begin{bmatrix} v_{1r} \\ v_{2r} \\ v_{3r} \end{bmatrix} \tag{2.12}$$

and

$$\mathbf{v}^s = \begin{bmatrix} v_{1s} \\ v_{2s} \\ v_{3s} \end{bmatrix} \tag{2.13}$$

To show how they are related, we notice that

$$\mathbf{v}^s = \begin{bmatrix} v_{1s} \\ v_{2s} \\ v_{3s} \end{bmatrix} = \begin{bmatrix} \mathbf{s}_1'\mathbf{v} \\ \mathbf{s}_2'\mathbf{v} \\ \mathbf{s}_3'\mathbf{v} \end{bmatrix} = \begin{bmatrix} \mathbf{s}_1' \\ \mathbf{s}_2' \\ \mathbf{s}_3' \end{bmatrix} \mathbf{v} = \begin{bmatrix} l_1 & m_1 & n_1 \\ l_2 & m_2 & n_2 \\ l_3 & m_3 & n_3 \end{bmatrix} \begin{bmatrix} \mathbf{r}_1' \\ \mathbf{r}_2' \\ \mathbf{r}_3' \end{bmatrix} \mathbf{v}$$

$$= \begin{bmatrix} l_1 & m_1 & n_1 \\ l_2 & m_2 & n_2 \\ l_3 & m_3 & n_3 \end{bmatrix} \begin{bmatrix} \mathbf{r}_1'\mathbf{v} \\ \mathbf{r}_2'\mathbf{v} \\ \mathbf{r}_3'\mathbf{v} \end{bmatrix} = \begin{bmatrix} l_1 & m_1 & n_1 \\ l_2 & m_2 & n_2 \\ l_3 & m_3 & n_3 \end{bmatrix} \mathbf{v}^r$$

The first equality follows from Eq. (2.13), the second from Eq. (2.11), the third from inner matrix product, the fourth from Eq. (2.5), the fifth from inner matrix product and the last from Eq. (2.12). This result can be summarized as

$$\mathbf{v}^s = \begin{bmatrix} l_1 & m_1 & n_1 \\ l_2 & m_2 & n_2 \\ l_3 & m_3 & n_3 \end{bmatrix} \mathbf{v}^r = \mathbf{C}_r^s \mathbf{v}^r \tag{2.14}$$

In the above equation we adopted the notation $\mathbf{C}_r^s$ to denote the DCM that transforms vector coordinates from frame ***R*** to frame ***S*** [3]. Eq. (2.14) is a fundamental equation upon which the rest of our development will be based. Had we switched the roles of **r** and **s** we would have arrived at

$$\mathbf{v}^r = \mathbf{C}_s^r \mathbf{v}^s \tag{2.15}$$

Now, from Eq. (2.14)

$$\mathbf{v}^r = \mathbf{C}_s^r \mathbf{v}^s = \mathbf{C}_s^r \mathbf{C}_r^s \mathbf{v}^r$$

Since **v** is an arbitrary vector, we can deduce that

$$\mathbf{C}_s^r \mathbf{C}_r^s = \mathbf{I}$$

and from Eq. (2.8)

$$\mathbf{C}_s^{r\prime} = \mathbf{C}_r^s \tag{2.16}$$

2.5 DCM Fixed Axis

The fixed axis is a vector that would not change its direction when operated on by the DCM. To find this vector, we first prove that the eigenvalue magnitudes of a unitary matrix are all 1's. Suppose that **e** and λ respectively, are an eigenvector and an eigenvalue pair of matrix **C**, i.e.

$$\mathbf{Ce} = \lambda\mathbf{e}$$

We will consider the possibility that an eigen pair, **e** and λ, of matrix **C** could be complex even if all its elements are real [2]. In such case their complex conjugates pair, $\mathbf{e}^*$ and λ^*, is also an eigen pair to **C**. With this in mind, we perform the Hermitian inner product defined in Chapter 1 as follows: *separately* transpose and conjugate the above equation to get

$$\mathbf{e}'\mathbf{C}' = \lambda\mathbf{e}'$$

and

$$\mathbf{C}^*\mathbf{e}^* = \lambda^*\mathbf{e}^*$$

Now, the product of both sides of the above two equations is

$$\mathbf{e}'\mathbf{C}'\mathbf{C}^*\mathbf{e}^* = \lambda\lambda^*\mathbf{e}'\mathbf{e}^*$$

Given that **C** is a real unitary matrix implies that $\mathbf{C}'\mathbf{C}^* = \mathbf{C}'\mathbf{C} = \mathbf{I}$ and hence simplifies the above equation to

$$\mathbf{e}'\mathbf{e}^* = \lambda\lambda^*\mathbf{e}'\mathbf{e}^* = |\lambda|^2\,\mathbf{e}'\mathbf{e}^*$$

which implies that

$$1 = |\lambda|^2$$

Since **C** has three eigenvalues and that complex eigenvalues occur in pairs, then **C** has at least one real eigenvalue which, with no loss of generality, equals 1. If **u** is the corresponding eigenvector, then

$$\mathbf{C}\,\mathbf{u} = \mathbf{u}$$

and it follows that **u** is the fixed axis. To determine **u**, we use the property that the eigenvectors of a matrix are the same as those of its inverse. Since **C** is unitary then

$$\mathbf{C}^{-1}\,\mathbf{u} = \mathbf{C}'\,\mathbf{u} = \mathbf{u}$$

Differencing the above two equations gives

$$(\mathbf{C} - \mathbf{C}')\mathbf{u} = \begin{bmatrix} 0 & c_{12} - c_{21} & c_{13} - c_{31} \\ c_{21} - c_{12} & 0 & c_{23} - c_{32} \\ c_{31} - c_{13} & c_{32} - c_{23} & 0 \end{bmatrix} \begin{bmatrix} u_1 \\ u_2 \\ u_3 \end{bmatrix} = 0$$

The skew symmetric matrix property, in Eq. (1.17), shows that the solution to this equation is

$$\begin{bmatrix} u_1 \\ u_2 \\ u_3 \end{bmatrix} = \begin{bmatrix} c_{32} - c_{23} \\ c_{13} - c_{31} \\ c_{21} - c_{12} \end{bmatrix}$$

This property will prove in the following that the DCM is a rotation transformation matrix.

2.6 The Rotation Matrix

Here we describe what a rotation matrix is and show its equivalence to a transformation matrix [4,5]. Figure 2.2 depicts a rotation axis, along the unit vector **u,** about which arbitrary vectors can rotate. In the figure, vector **a** (represented as $\overrightarrow{ma}$) intersects **u** at m. As **a** rotates about **u**, point a will trace a circle centered at o. Vector $\overrightarrow{ob}$ is constructed to lie in the circle plane and at right angle to **a**. Let **a** make an arbitrary angular rotation ϕ about **u** to move to **d**. Our goal is to describe **d** as a function of **a**, **u** and ϕ. One can see that **a** is the sum of 2 orthogonal vectors: $\overrightarrow{mo}$, the projection of **a** along **u**, and $\overrightarrow{oa}$, the projection of **a** on the circle plane.

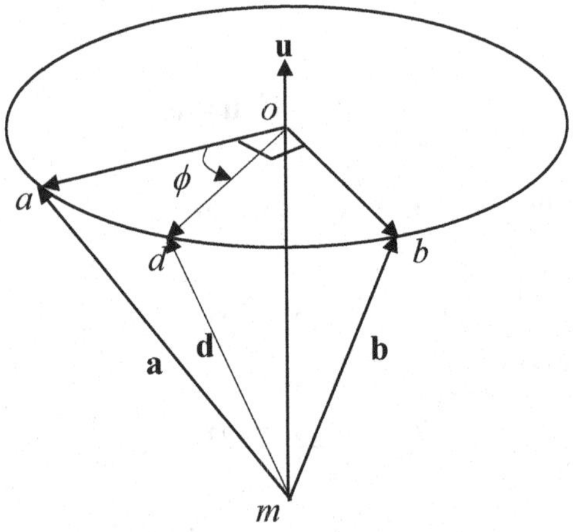

Figure 2.2 Rotating Vector **a** about Axis **u** to New Location **d**

In the following we will use vectors and column matrices interchangeably. With that in mind we can write

$$\overrightarrow{mo} = (\mathbf{u}'\mathbf{a})\,\mathbf{u} = \mathbf{u}\mathbf{u}'\mathbf{a}$$

from which we get

$$\overrightarrow{oa} = \overrightarrow{ma} - \overrightarrow{mo} = \mathbf{a} - \mathbf{u}\mathbf{u}'\mathbf{a} = (\mathbf{I}\text{-}\mathbf{u}\mathbf{u}')\mathbf{a}$$

Since $\overrightarrow{ob}$ is orthogonal to the plane of **u** and $\overrightarrow{oa}$, and that $|\overrightarrow{ob}| = |\overrightarrow{oa}|$ (two radii in the same circle), then

$$\overrightarrow{ob} = \mathbf{u} \times \overrightarrow{oa}$$

Converting the cross product into matrix form in the above equation using Eq. (1.14) gives

$$\overrightarrow{ob} = \mathbf{u} \times \overrightarrow{oa} = \tilde{\mathbf{u}}\, \overrightarrow{oa}$$

Substituting for $\overrightarrow{oa}$ in the above equation and using Eq. (1.17) we get

$$\overrightarrow{ob} = \tilde{\mathbf{u}}\ (\mathbf{I} - \mathbf{uu}')\mathbf{a} = \ \tilde{\mathbf{u}}\mathbf{a}$$

Let $c = \cos\phi$ and $s = \sin\phi$. Since the projections of $\overrightarrow{od}$ on the orthogonal vectors, $\overrightarrow{oa}$ and $\overrightarrow{ob}$, are $c\ \overrightarrow{oa}$ and $s\ \overrightarrow{ob}$, respectively, then

$$\overrightarrow{od} = c\ \overrightarrow{oa} + s\ \overrightarrow{ob}$$

and therefore the rotated vector, **d,** is

$$\begin{aligned} \mathbf{d} &= \overrightarrow{md} = \overrightarrow{mo} + \overrightarrow{od} \\ &= \overrightarrow{mo} + c\ \overrightarrow{oa} + s\ \overrightarrow{ob} \end{aligned}$$

Substituting for $\overrightarrow{mo}, \overrightarrow{oa}$ and $\overrightarrow{ob}$ in terms of their matrix forms gives

$$\mathbf{d} = (\mathbf{uu}')\mathbf{a} + c(\mathbf{I} - \mathbf{uu}')\mathbf{a} + s\tilde{\mathbf{u}}\mathbf{a}$$

Collecting terms in the above equation yields

$$\begin{aligned}\mathbf{d} &= \left[\mathbf{uu}' + c(\mathbf{I} - \mathbf{uu}') + s\tilde{\mathbf{u}}\right]\mathbf{a} \\ &= \left[\, c\mathbf{I} + (1-c)\mathbf{uu}' + s\tilde{\mathbf{u}}\right]\mathbf{a}\end{aligned}$$

Therefore

$$\mathbf{C} = c\mathbf{I} + (1-c)\mathbf{uu}' + s\tilde{\mathbf{u}} \tag{2.17}$$

is the matrix that rotates the vector **a** about **u** by an angle ϕ to its new location **d**. Eq. (2.17) is known as the Rodrigues formula [4-6]. Employing the properties of the skew symmetric matrix, it is straightforward to prove that

$$\mathbf{CC}' = \mathbf{I}$$

Thus **C** is a unitary matrix and hence is a transformation matrix. Moreover, operating on **u** gives

$$\mathbf{Cu} = \mathbf{u}$$

This shows that the fixed axis that was found in section 2.4 is also a rotation axis. We now conclude that a DCM, a transformation matrix, a unitary matrix and a rotation matrix are different faces of the same object. It is evident from Eq. (2.17) that the rotation axis, **u**, and angle of rotation, ϕ, are all what is needed to construct the DCM.

With reference to Fig. 2.2, we define the rotation vector, $\overline{\boldsymbol{\varphi}}$, to be that of magnitude, ϕ rad, and points along the unit vector **u**, i.e.

$$\overline{\boldsymbol{\varphi}} = \phi\,\mathbf{u} \tag{2.18}$$

The rotation vector comprises the minimal information needed to construct the DCM.

Finally, we observe that since the product of unitary matrices is also unitary, then the product of DCM's is another DCM. Thus we can construct the DCM if we know the sequence of rotations from one frame to another.

2.7 Inner and Outer Transformation Matrices

Even though rotation and transformation matrices share the same properties, they are not the same. The difference between them can be illustrated graphically by the example in Fig. 2.3. Fig. 2.3a depicts a two-dimensional vector **v** in the *x-y* plane. Fig. 2.3b depicts the same vector (stationary) while the *x-y* frame rotates an angle ϕ about the *z*-axis. Fig. 2.3c depicts the same vector as it rotates an angle ϕ about the *z*-axis and the frame is stationary. As will be evident shortly, we call the matrix that rotates the frame, as in Fig. 2.3b, an outer transformation matrix, and the matrix that rotates the vector, as in Fig. 2.3c, an inner transformation matrix. Had the vector in Fig. 2.3a rotated an angle $-\phi$ (while the frame is fixed) it would have arrived at the same relative position to the frame as in Fig. 2.3b.

These rotations might be illustrated further by visualizing a movie camera (to which attached a frame of axes) that stands at the center of a circle, and on whose circumference is a train track. A train (tip of a vector) that moves counter clockwise will appear to the camera exactly the same as when the train is stationary and the camera rotates clockwise. We may conclude from the above discussion that an outer transformation matrix that corresponds to a frame rotation of ϕ about an axis of rotation **u** is exactly the same as an inner transformation matrix that rotates a vector an angle $-\phi$ about the same axis of rotation.

We extend the above to a relevant example in which the concept of inner and outer rotations is more apparent. Here, we will be dealing with two frames, one is attached to the body of an aircraft and the second (called the navigation frame) is located at some point on Earth but not attached to it. In this example, these two frames can rotate relative to each other independently. For instance, due to aircraft maneuvers, the body frame rotates relative to the Earth. Conversely, the navigation frame rotates when the craft moves from one place on Earth to another, say from Los Angeles to Paris. We track the rotation from the body to the navigation frame by the DCM, $\mathbf{C}_b^n$. This DCM is updated by incremental changes when either of these two frames rotates.

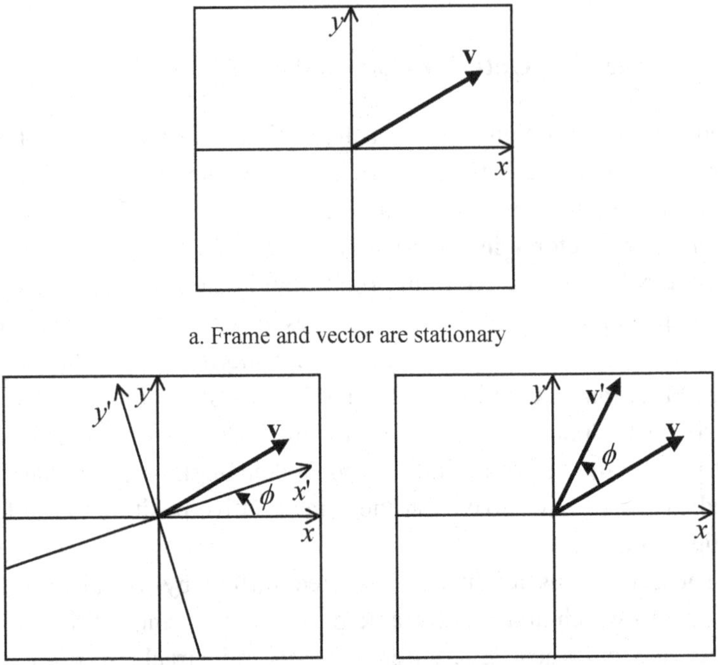

a. Frame and vector are stationary

b. Outer Transformation: frame rotates an angle ϕ while vector is stationary

c. Inner Transformation: vector rotates an angle ϕ while frame is stationary

Figure 2.3 Relative rotations of frame and vector

For navigation purposes, a triad of orthogonal accelerometers is fixed along the axes of the body frame. The readings of the accelerometer set are the components of $\mathbf{a}^b$, the acceleration vector that is represented in the body frame. In the navigation frame this acceleration vector is given by

$$\mathbf{a}^n = \mathbf{C}_b^n \mathbf{a}^b \tag{2.19}$$

First, we discuss the case of body frame rotation from b a new location denoted by $\bar{b}$. We denote the incremental DCM from b to $\bar{b}$ by $\mathbf{C}_{\bar{b}}^{b}$, and the new acceleration vector (the accelerometer measurements in the new frame) by $\mathbf{a}^{\bar{b}}$. The acceleration vector in the navigation frame becomes

$$
\begin{aligned}
\mathbf{a}^{n} &= \mathbf{C}_{\bar{b}}^{n}\mathbf{a}^{\bar{b}} \\
&= \mathbf{C}_{b}^{n}\mathbf{C}_{\bar{b}}^{b}\mathbf{a}^{\bar{b}}
\end{aligned}
\tag{2.20}
$$

In the above equation, notice the location of the incremental DCM, $\mathbf{C}_{\bar{b}}^{b}$. Since it is a case of vector (accelerometer) rotation, the incremental DCM is an inner one and its sense of rotation should be opposite to that of the vector rotation as indicated by the sub and superscript of $\mathbf{C}_{\bar{b}}^{b}$.

Next, we address the case in which the body frame is unchanged, but the navigation frame rotates to a new location denoted by $\bar{n}$ for which the DCM from n to $\bar{n}$ is $\mathbf{C}_{n}^{\bar{n}}$. The acceleration vector in the new frame becomes

$$
\begin{aligned}
\mathbf{a}^{\bar{n}} &= \mathbf{C}_{b}^{\bar{n}}\mathbf{a}^{b} \\
&= \mathbf{C}_{n}^{\bar{n}}\mathbf{C}_{b}^{n}\mathbf{a}^{b}
\end{aligned}
\tag{2.21}
$$

Now notice the location of the incremental DCM, $\mathbf{C}_{n}^{\bar{n}}$. Since it is a case of frame rotation, the incremental DCM is an outer one and its sense of rotation should be the same as that of the frame rotation.

When we have both body and navigation frame rotations the acceleration vector is given by

$$
\mathbf{a}^{\bar{n}} = \mathbf{C}_{n}^{\bar{n}}\mathbf{C}_{b}^{n}\mathbf{C}_{\bar{b}}^{b}\mathbf{a}^{\bar{b}} \tag{2.22}
$$

and in this situation we have both an inner and an outer incremental DCMs.

We may view of Eq. (2.17) as a DCM that corresponds to rotating a vector an angle ϕ. From the above discussion if a DCM for which the frame rotates an angle ϕ, the equation of the DCM would have changed from Eq. (2.17) to

$$\mathbf{C} = \cos(-\phi)\mathbf{I} + [1\text{-}\cos(-\phi)]\mathbf{u}\mathbf{u}' + \sin(-\phi)\tilde{\mathbf{u}}$$

or

$$\mathbf{C} = c\mathbf{I} + (1-c)\mathbf{u}\mathbf{u}' - s\tilde{\mathbf{u}} \tag{2.23}$$

where $c = \cos\phi$ and $s = \sin\phi$.

2.8 The Quaternion

In the above discussion we have observed that the rotation axis, **u**, and the rotation angle, ϕ, are all we need to construct the DCM. Yet, there is another way to form the DCM.

Let

$$c' = \cos(\phi/2) \tag{2.24a}$$

and

$$s' = \sin(\phi/2) \tag{2.24b}$$

Using the trigonometric identities

$$c = \cos(\phi) = \cos^2(\phi/2) - \sin^2(\phi/2) = c'^2 - s'^2$$

and

$$s = \sin\phi = 2\sin\frac{\phi}{2}\cos\frac{\phi}{2} = 2\,s'c'$$

to substitute for c and s in $\mathbf{C}$ in Eq. (2.23) yields

$$\mathbf{C} = (c'^2 - s'^2)\mathbf{I} + 2s'^2\,\mathbf{u}\mathbf{u}' - 2s'c'\tilde{\mathbf{u}}$$

Let us define q_0 and $\mathbf{q}$ to be

$$q_0 = c' \text{ and } \mathbf{q} = s'\mathbf{u} \tag{2.25}$$

which implies that

$$\mathbf{q}'\mathbf{q} = s'^2\,\mathbf{u}'\mathbf{u} = s'^2 \tag{2.26}$$

Substituting for c' and s' from Eqs. (2.25) and (2.26) in $\mathbf{C}$ yields

$$\mathbf{C} = (q_0^2 - \mathbf{q}'\mathbf{q})\mathbf{I} + 2\mathbf{q}\mathbf{q}' - 2q_0\tilde{\mathbf{q}} \tag{2.27}$$

The above development was just a back door for introducing the quaternion, q_0 and $\mathbf{q}$, which will be discussed in more detail in the following chapter. It shows that the quaternion, formed from the rotation axis and the rotation angle given in Eq. (2.18), can be used to form the DCM.

This chapter was dedicated to introducing the concept of the rotation vector and the coordinate transformation. Both play a central role for the development of the inertial navigation equations. In the following chapter we explore alternative algorithms for transforming vectors between different frames.

References

1. H. Schneider, G. P. Barker, Matrices and Linear Algebra, Holt Rinehart and Winston Inc, New York, 1968.
2. C. G. Cullen, Matrices and Linear Transformations, Addison-Wesley Co., Reading, MA, 1972.
3. Britting, K. R, Inertial Navigation System Analysis, Wiley Interscience, NY, New York, 1971.
4. Rodrigues' Rotation Formula -- from Wolfram MathWorld
5. http://planetmath.org/encyclopedia/ProofOfRodriguesRotationFormula.html
6. H. Goldstein, Classical Mechanics, Addison-Wesley Co., Reading, MA, 1981.

Chapter 3

Forms of the Transformation Matrix

3.1 Introduction

As discussed in Chapter 2, the direction cosine matrix (DCM) is the tool that transforms vectors from one coordinate frame to another. The elements of this matrix are the direction cosines of the principal axes of one frame to the other. However to construct the DCM from the direction cosines is not only unattractive, but also inconvenient. One reason is that the orthonormality property is lost and with it the length and orientation of the transformed vector. To preserve this property one can use the rotation matrix to construct the DCM. That is, to find the angle of rotation and the axis around which the moving frame has rotated. Two very closely related methods fall under this category: the rotation vector and the quaternion. Another elegant approach is to use the Euler angles. This approach is based on using a sequence of simple rotations about independent axes rather than one single rotation; the DCM is then the product of simple rotation matrices. It will be shown that rotations about the principal (x, y, z) axes take very simple and standard formats. All we need is to know the amount of rotation associated with each axis to form these DCMs.

This chapter has two main objectives. The first is to introduce the quaternion and show how to use it to construct the DCM. We will describe the means for converting from one form of transformation to another, e.g. converting from Euler angles to quaternions. With our inertial navigation system in mind, we desire to convert vectors to frames that continually change with time. Thus our second goal is to derive the differential equations that govern the dynamics of the DCM. We start by introducing the DCMs that perform simple rotations.

3.2 Simple Frame Rotations

The simple frame rotations are those that rotate about the x, y and z-axes. We will be using Eq. (2.17) to construct the respective DCM. Herein, we adopt these shorthand notations: cϕ or c(ϕ) to mean cos(ϕ), and likewise sϕ or s(ϕ) to mean sin(ϕ). When there is no confusion, the parenthesless shorthand will be adopted. Also we use $\mathbf{C}_x(\phi)$ to denote a DCM that performs an angular rotation ϕ about the x-axis.

x-axis: the DCM that rotates an angle ϕ about $\mathbf{u} = [1 \quad 0 \quad 0]'$ is

$$\mathbf{C}_x(\phi) = \begin{bmatrix} 1 & 0 & 0 \\ 0 & \mathrm{c}\phi & \mathrm{s}\phi \\ 0 & -\mathrm{s}\phi & \mathrm{c}\phi \end{bmatrix} \tag{3.1}$$

y-axis: the DCM that rotates an angle θ about $\mathbf{u} = [0 \quad 1 \quad 0]'$ is

$$\mathbf{C}_y(\theta) = \begin{bmatrix} \mathrm{c}\theta & 0 & -\mathrm{s}\theta \\ 0 & 1 & 0 \\ \mathrm{s}\theta & 0 & \mathrm{c}\theta \end{bmatrix} \tag{3.2}$$

z-axis: the DCM that rotates an angle ψ about $\mathbf{u} = [0 \quad 0 \quad 1]'$ is

$$\mathbf{C}_z(\psi) = \begin{bmatrix} \mathrm{c}\psi & \mathrm{s}\psi & 0 \\ -\mathrm{s}\psi & \mathrm{c}\psi & 0 \\ 0 & 0 & 1 \end{bmatrix} \tag{3.3}$$

3.3 Euler Angles

As mentioned earlier, the Euler angles approach forms the DCM through a sequence of simple rotations. Although there is no unique sequence for forming the Euler angles, there are widely acceptable sets for specific applications. One of them is the rotation sequence about aircraft body axes. In this frame, the axes are defined so that the x-axis points to the nose along the symmetric longitudinal axis of the craft, the y-axis is orthogonal to the x-axis along the right wing, and the z-axis is perpendicular to the x-y plane and points downwards. In this configuration it is assumed that when the craft is level, the x-y plane will be horizontal and hence the z-axis will be vertical to the ground. As shown in Fig 3.1, the Euler sequence is performed such that the reference frame first rotates an angle ψ (yaw) about the z-axis to form $x'y'z'$ frame. Since the rotation is about the z-axis, then z and z' will be the same. Second, the new frame is rotated an angle θ (pitch) about the y'-axis to form $x''y''z''$. Finally this frame is rotated an angle ϕ (roll) about its x''-axis to form $x'''y'''z'''$ frame. We observe that the DCMs that describe these rotations are those special matrices given in the previous section.

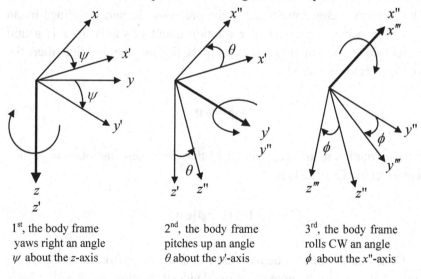

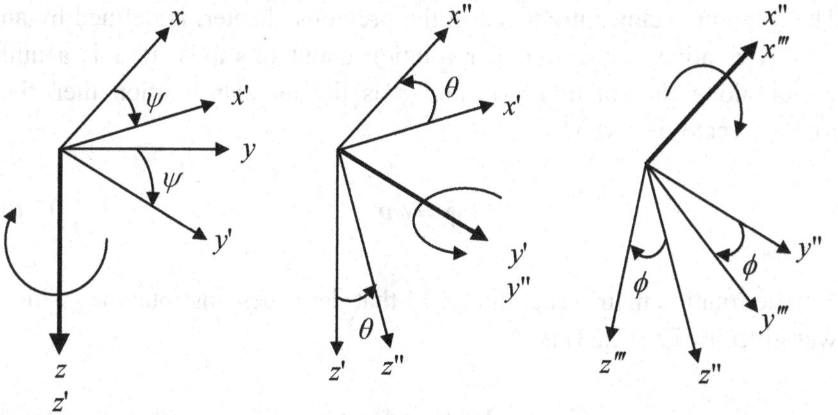

1st, the body frame yaws right an angle ψ about the z-axis

2nd, the body frame pitches up an angle θ about the y'-axis

3rd, the body frame rolls CW an angle ϕ about the x''-axis

Figure 3.1 Euler Angles Sequence

The DCM product of the respective rotations about the z, y' and x'' axes is then

$$\mathbf{C}_n^b = \mathbf{C}_{x''}(\phi)\mathbf{C}_{y'}(\theta)\mathbf{C}_z(\psi) \tag{3.4}$$

Using Eqs. (3.1) - (3.3), one can expand the above equation into

$$\mathbf{C}_n^b = \begin{bmatrix} \mathrm{c}\theta\,\mathrm{c}\psi & \mathrm{c}\theta\,\mathrm{s}\psi & -\mathrm{s}\theta \\ \mathrm{s}\phi\,\mathrm{s}\theta\,\mathrm{c}\psi - \mathrm{c}\phi\,\mathrm{s}\psi & \mathrm{s}\phi\,\mathrm{s}\theta\,\mathrm{s}\psi + \mathrm{c}\phi\,\mathrm{c}\psi & \mathrm{s}\phi\,\mathrm{c}\theta \\ \mathrm{c}\phi\,\mathrm{s}\theta\,\mathrm{c}\psi + \mathrm{s}\phi\,\mathrm{s}\psi & \mathrm{c}\phi\,\mathrm{s}\theta\,\mathrm{s}\psi - \mathrm{s}\phi\,\mathrm{c}\psi & \mathrm{c}\varphi\,\mathrm{c}\theta \end{bmatrix} \tag{3.5}$$

We now discuss the rotation vector and the quaternion which are two powerful methods for expressing transformation matrices.

3.4 Rotation Vector

The rotation vector, introduced in the previous chapter, is defined by an axis of rotation and an angular rotation about this axis. If $\mathbf{u}$ is a unit vector along the rotation axis and ϕ is the angular rotation then the rotation vector is given by

$$\bar{\boldsymbol{\varphi}} = \phi\mathbf{u} \tag{3.6}$$

The rotation matrix (i.e. the DCM that describes this rotation), which was given by Eq. (2.23) is

$$\mathbf{C} = \mathrm{c}\phi\,\mathbf{I} + (1\text{-}\mathrm{c}\phi)\mathbf{uu}' - \mathrm{s}\phi\,\tilde{\mathbf{u}} \tag{3.7}$$

Conversely, one can deduce the rotation vector from a given DCM. However it should be noted that the deduced rotation angle would be in terms of its sine and cosine rather than the explicit rotation angle.

3.5 Quaternion

One may view the quaternion as an alternative form to a rotation vector in which it is associated with the sine and the cosine of the rotation angle rather than the actual rotation angle associated with the rotation vector.

A quaternion, as its name imply, is a four-scalar set. It is represented by a 4-dimensional vector

$$\mathbf{Q} = [q_0,\ q_1,\ q_2,\ q_3] \tag{3.8}$$

or equivalently by a scalar and 3-dimensional vector $\mathbf{Q} = \{q_0, \mathbf{q}\}$ with $\mathbf{q} = [\, q_1,\ q_2,\ q_3\,]$. Both forms are commonly used, as the former expression fits well for matrix implementations, while the later is useful for formula derivation and shorthand notations. A brief introduction to quaternion algebra is presented in Appendix B. We recall that the identity quaternion is

$$\mathbf{Q}_\mathbf{I} = [1,0,0,0]$$

The product of the two quaternions $\mathbf{P} = \{p_0, \mathbf{p}\}$ and $\mathbf{Q} = \{q_0, \mathbf{q}\}$ is also a quaternion given by

$$\mathbf{PQ} = \{p_0 q_0 - (\mathbf{pq}),\ p_0\,\mathbf{q} + q_0\,\mathbf{p} + \mathbf{p} \times \mathbf{q}\} \tag{3.9}$$

The norm of the quaternion is

$$\text{norm}(\mathbf{Q}) = \sqrt{q_0^2 + q_1^2 + q_2^2 + q_3^2} \tag{3.10}$$

The inverse of a quaternion **Q,** denoted it by $\mathbf{Q}^{\#}$, have the property

$$\mathbf{QQ}^{\#} = \mathbf{Q}^{\#}\mathbf{Q} = \mathbf{Q}_\mathbf{I}$$

From Eq. (3.9) one can verify that the inverse of a quaternion **Q** is given by

$$\mathbf{Q}^{\#} = \{q_0, -\mathbf{q}\} / \operatorname{norm}^2(\mathbf{Q}) \tag{3.11}$$

A normalized quaternion is one whose norm equals 1, that is if $\mathbf{Q} = \{q_0, \mathbf{q}\}$ then

$$q_0^2 + q_1^2 + q_2^2 + q_3^2 = 1 \tag{3.12}$$

and its inverse is given by

$$\mathbf{Q}^{\#} = \{q_0, -\mathbf{q}\} \tag{3.13}$$

The class of normalized quaternions plays a central role in matrix transformations as seen in the following.

Lemma 3.1

Suppose a three-dimensional vector $\mathbf{v} = [v_1, v_2, v_3]$ is cast into a quaternion $\mathbf{V} = \{0, \mathbf{v}\}$ and $\mathbf{Q} = \{q_0, \mathbf{q}\}$ is a normalized quaternion then

$$\mathbf{Q}^{\#}\mathbf{V}\mathbf{Q} = \{0, \mathbf{C}\mathbf{v}\} \tag{3.14}$$

where

$$\mathbf{C} = (q_0^2 - \mathbf{q}'\mathbf{q})\,\mathbf{I} + 2\mathbf{q}\mathbf{q}' - 2q_0\tilde{\mathbf{q}} \tag{3.15}$$

Note that, in Eqs. (3.14) and (3.15), **v** and **q** play the column array roles that derive from their respective vectors.

Proof: Perform the quaternion product

$$\mathbf{Q}^{\#}\mathbf{V} = \{q_0,\text{-}\mathbf{q}\}\{0,\mathbf{v}\} = \{(\mathbf{q}.\mathbf{v}), q_0\mathbf{v} - \mathbf{q}\times\mathbf{v}\}$$

Post multiply by **Q**

$$\mathbf{Q}^{\#}\mathbf{VQ} = \{(\mathbf{q}.\mathbf{v}), q_0\mathbf{v}\text{-}\mathbf{q}\times\mathbf{v}\}\{q_0,\mathbf{q}\}$$

Expand the above quaternion product, using Eq. (3.9), to get

$$\mathbf{Q}^{\#}\mathbf{VQ} = \begin{Bmatrix} q_0(\mathbf{q}.\mathbf{v})\text{-}q_0(\mathbf{q}.\mathbf{v}) + (\mathbf{q}.\mathbf{q}\times\mathbf{v}) \\ ,(\mathbf{q}.\mathbf{v})\mathbf{q} + q_0(q_0\mathbf{v}\text{-}\mathbf{q}\times\mathbf{v}) + (q_0\mathbf{v}\text{-}(\mathbf{q}\times\mathbf{v}))\times\mathbf{q} \end{Bmatrix}$$

Since

$$\mathbf{q}.(\mathbf{q}\times\mathbf{v}) = 0\,,\ \mathbf{v}\times\mathbf{q} = \text{-}\mathbf{q}\times\mathbf{v} \text{ and } (\mathbf{q}\times\mathbf{v})\times\mathbf{q} = (\mathbf{q}.\mathbf{q})\mathbf{v} - (\mathbf{v}.\mathbf{q})\mathbf{q}$$

then

$$\mathbf{Q}^{\#}\mathbf{VQ} = \left\{0,(\mathbf{q}.\mathbf{v})\mathbf{q} + q_0^2\mathbf{v}\text{-}2q_0\ \mathbf{q}\times\mathbf{v}\text{-}(\mathbf{q}.\mathbf{q})\,\mathbf{v} + (\mathbf{v}.\mathbf{q})\mathbf{q}\right\}$$

Simplifying the vector part of the above product

$$\begin{aligned} \bar{\mathbf{v}} &= (\mathbf{q}.\mathbf{v})\mathbf{q} + q_0^2\mathbf{v}\text{-}2q_0\ \mathbf{q}\times\mathbf{v}\text{-}(\mathbf{q}.\mathbf{q})\,\mathbf{v} + (\mathbf{v}.\mathbf{q})\mathbf{q} \\ &= (q_0^2\text{-}(\mathbf{q}.\mathbf{q}))\,\mathbf{v} + 2(\mathbf{q}.\mathbf{v})\mathbf{q}\text{-}2q_0\mathbf{q}\times\mathbf{v} \end{aligned}$$

When the above vector is converted to matrix notation it becomes

$$\bar{\mathbf{v}} = (q_0^2 \text{-} \mathbf{q}'\mathbf{q})\mathbf{v} + 2\mathbf{q}(\mathbf{q}'\mathbf{v}) - 2q_0\,\tilde{\mathbf{q}}\mathbf{v} = \mathbf{C}\mathbf{v}$$

where

$$\mathbf{C} = (q_0^2 \text{-} \mathbf{q}'\mathbf{q}) + 2\mathbf{q}\mathbf{q}' \text{-} 2q_0\tilde{\mathbf{q}}$$

Q.E.D.

Lemma 3.2

The quaternion that describes a frame rotation of angle ϕ about the unit vector $\mathbf{u}$ is given by

$$\mathbf{Q} = \{q_0, \mathbf{q}\} = \{\mathrm{c}\left(\frac{\phi}{2}\right), \mathrm{s}\left(\frac{\phi}{2}\right)\mathbf{u}\} \tag{3.16}$$

This quaternion is exactly the same as that given in Eq. (2.25). By using Eq. (2.18), this quaternion can be equivalently represented by

$$\mathbf{Q} = \{q_0, \mathbf{q}\} = \{\mathrm{c}\left(\frac{\phi}{2}\right), \mathrm{s}\left(\frac{\phi}{2}\right)\frac{\bar{\boldsymbol{\varphi}}}{\phi}\} \tag{3.17}$$

The above illuminates the relevance of the rotation vector to the quaternion. Further, it provides two alternatives for transforming a vector using the quaternion:

1. Perform the product $\mathbf{Q}^{\#}\mathbf{V}\mathbf{Q}$
2. Convert the quaternion to a DCM as in Eq. (2.27) and then perform the transformation.

In section 3.1 we have described the simple rotation DCMs about the principal axes. In the following we describe their quaternion counterparts: those quaternions that when operated on a vector yield the same transformations.

3.6 Simple Quaternions

In the following we shall use the shorthand notation of $\mathbf{Q}_x(\phi)$ to describe a quaternion that causes a rotation of angle ϕ about the x-axis.

1. x-axis: Eq. (3.16) show that the quaternion that rotates an angle ϕ about $\mathbf{u} = [1 \quad 0 \quad 0]'$ is

$$\mathbf{Q}_x(\phi) = \{c\frac{\phi}{2}, [s\frac{\phi}{2}, 0, 0]\} \tag{3.18}$$

Using Eq. (2.27) we see that it corresponds to the DCM in Eq. (3.1).

$$\mathbf{C}_x(\phi) = \begin{bmatrix} 1 & 0 & 0 \\ 0 & c\phi & s\phi \\ 0 & -s\phi & c\phi \end{bmatrix}$$

Thus this quaternion will perform a rotation ϕ about the x-axis

2. y-axis: Eq. (3.16) show that the quaternion that rotates an angle θ about $\mathbf{u} = [0 \quad 1 \quad 0]'$ is

$$\mathbf{Q}_y(\theta) = \{c\frac{\theta}{2}, [0, s\frac{\theta}{2}, 0]\} \tag{3.19}$$

which corresponds to the DCM given in Eq. (3.2)

$$\mathbf{C}_y(\theta) = \begin{bmatrix} c\theta & 0 & -s\theta \\ 0 & 1 & 0 \\ s\theta & 0 & c\theta \end{bmatrix}$$

3. z-axis: Eq. (3.16) show that the quaternion that rotates an angle ψ about $\mathbf{u} = [0 \quad 0 \quad 1]'$ is

$$\mathbf{Q}_z(\psi) = \{c\frac{\psi}{2}, [0,0,s\frac{\psi}{2}]\} \tag{3.20}$$

which corresponds to the DCM given in Eq. (3.3)

$$\mathbf{C}_z(\psi) = \begin{bmatrix} c\psi & s\psi & 0 \\ -s\psi & c\psi & 0 \\ 0 & 0 & 1 \end{bmatrix}$$

Therefore, the quaternion that performs the Euler angles sequence is given by

$$\mathbf{Q}_n^b = \mathbf{Q}_z(\psi)\,\mathbf{Q}_y(\theta)\mathbf{Q}_x(\phi) \tag{3.21}$$

When expanded it becomes

$$\mathbf{Q}_n^b = \begin{bmatrix} c\frac{\psi}{2}c\frac{\theta}{2}c\frac{\phi}{2} + s\frac{\psi}{2}s\frac{\theta}{2}s\frac{\phi}{2} \\ c\frac{\psi}{2}c\frac{\theta}{2}s\frac{\phi}{2} - s\frac{\psi}{2}s\frac{\theta}{2}c\frac{\phi}{2} \\ c\frac{\psi}{2}s\frac{\theta}{2}c\frac{\phi}{2} + s\frac{\psi}{2}c\frac{\theta}{2}s\frac{\phi}{2} \\ s\frac{\psi}{2}c\frac{\theta}{2}c\frac{\phi}{2} - c\frac{\psi}{2}s\frac{\theta}{2}s\frac{\phi}{2} \end{bmatrix} \tag{3.22}$$

3.7 Conversion between Forms

From the above we have four forms for expressing coordinate transformations: the DCM, the Euler angles, the rotation vector and the quaternion. Next we discuss and summarize the conversion between the different forms. Due to the close relation between the quaternion and the rotation vector, conversion to/from the rotation vector is not addressed. If needed, conversion to/or from the quaternion will be carried and then followed by using Eq. (3.17) or its inverse.

3.7.1 Conversion between DCM and Euler

Given the Euler angles, their sines and cosines can be computed and substituted in Eq. (3.5) to give the DCM. Conversely, when the DCM is known then Eq. (3.5) gives

$$\begin{aligned} \phi &= \arctan(\mathbf{C}_{23},\mathbf{C}_{33}) \\ \theta &= \operatorname{arcsine}(-\mathbf{C}_{13}) \\ \psi &= \arctan(\mathbf{C}_{12},\mathbf{C}_{11}) \end{aligned} \tag{3.23}$$

where $\mathbf{C}_{ij}$ is the (i^{th} row, j^{th} column) element of the matrix **C**.

3.7.2 Conversion between DCM and Quaternion

Given the quaternion, its elements can be substituted in Eq. (3.15) to generate the DCM. Expanding yields

$$\mathbf{C} = \begin{bmatrix} q_0^2+q_1^2-q_2^2-q_3^2 & 2(q_1q_2+q_0q_3) & 2(q_1q_3-q_0q_2) \\ 2(q_1q_2-q_0q_3) & q_0^2-q_1^2+q_2^2-q_3^2 & 2(q_2q_3+q_0q_1) \\ 2(q_1q_3+q_0q_2) & 2(q_2q_3-q_0q_1) & q_0^2-q_1^2-q_2^2+q_3^2 \end{bmatrix} \tag{3.24}$$

Now we compute **Q** when **C** is given. The trace of **C** (the sum of its diagonal terms) is given by

$$t = 3q_0^2 - (q_1^2 + q_2^2 + q_3^2)$$

In view of Eq. (3.12) the above becomes

$$t = 4q_0^2 - 1$$

from which

$$q_0 = \tfrac{1}{2}\sqrt{t+1} \tag{3.25}$$

Selecting (-180,180) to be the range of a rotation angle implies that the cosine of one half of any value in this range will always be positive and hence q_0, from Eq. (3.16), will be positive.

Continuing with Eq. (3.15) it can be seen that

$$\mathbf{C}' - \mathbf{C} = 4q_0\tilde{\mathbf{q}}$$

from which

$$\tilde{\mathbf{q}} = \frac{1}{4q_0}(\mathbf{C}' - \mathbf{C}) \tag{3.26}$$

As mentioned before, to convert a DCM to a rotation vector we convert it to a quaternion as given above then to a rotation vector. From Eq. (3.16) the rotation vector parameters are given by the two equations

$$\mathbf{u} = \frac{1}{|\mathbf{q}|}\mathbf{q}$$

$$\phi = 2\cos^{-1}(q_0)$$

3.7.3 Conversion between Euler Angles and Quaternion

Knowing the Euler angles implies that the sines and cosines of one half of their values can be computed and substituted in Eq. (3.22) to get the quaternion. On the other hand, computing the Euler angles from the quaternion is done in two steps: converting the quaternion to a DCM and then using Eq. (3.23) to determine the Euler angles. However we only construct the DCM elements needed to compute the angles. From Sections (3.4.1) and (3.4.2)

$$\begin{aligned} \phi &= \arctan(\mathbf{C}_{23}, \mathbf{C}_{33}) \\ &= \arctan(2(q_2 q_3 + q_0 q_1), q_0^2 - q_1^2 - q_2^2 + q_3^2) \end{aligned} \tag{3.27}$$

$$\begin{aligned} \theta &= \text{arcsine}(-\mathbf{C}_{13}) \\ &= \text{arcsine}(-2(q_1 q_3 - q_0 q_2)) \end{aligned} \tag{3.28}$$

$$\begin{aligned} \psi &= \arctan(\mathbf{C}_{12}, \mathbf{C}_{11}) \\ &= \arctan(2(q_1 q_2 + q_0 q_3), q_0^2 + q_1^2 - q_2^2 - q_3^2) \end{aligned} \tag{3.29}$$

Appendix C provides the Simulink® implementation of all the conversions. Now that we have the means for forming the DCM we proceed next to derive the differential equations that govern its change with respect to time.

3.8 Dynamics of the Transformation Matrix

Transformation matrices are perceived as means of transforming a vector from one frame to another. Thus far, both the vector and the DCM have been assumed static; that is both are stationary. But one might ask, what if one frame is continuously changing its direction, can we still construct the transformation matrix? Specifically we consider the case when one frame is rotating relative to another frame at some angular velocity.

We would like to determine how the transformation matrix will vary with time; that is, to derive its derivative with respect to time. This derivation is extended to the DCM, quaternion, rotation vector and Euler angles.

3.8.1 DCM Differential Equation

Consider the case of two initially coincident frames ***a*** and ***b*** and in which frame ***b*** rotates relative to fixed frame ***a***. Let's adopt this notation: at time *t*, the transformation matrix from ***b*** to ***a*** will be given by

$$\mathbf{C}_b^a(t) = \mathbf{C}_{b(t)}^a \tag{3.30}$$

Therefore the transformation matrix at time $t+dt$, will be

$$\mathbf{C}_b^a(t+dt) = \mathbf{C}_{b(t+dt)}^a = \mathbf{C}_{b(t)}^a \, \mathbf{C}_{b(t+dt)}^{b(t)} = \mathbf{C}_b^a(t) \, \mathbf{C}_{b(t+dt)}^{b(t)} \tag{3.31}$$

With reference to Fig. 2.2, we define the following terms: $\mathbf{u}$ is a unit vector along the axis about which vectors rotate and w is the angular rate of the vector frame about $\mathbf{u}$. The instantaneous angular rate vector, $\boldsymbol{\omega}$, is then

$$\boldsymbol{\omega} = w \, \mathbf{u} \tag{3.32}$$

Therefore in the time interval dt, frame ***b*** will rotate an angle given by

$$\phi = w \, dt \tag{3.33}$$

From the above and from Eq. (3.15), the DCM that govern this rotation is given by

$$\mathbf{C}_{b(t+dt)}^{b(t)} = \mathrm{c}\phi \, \mathbf{I} + (1 - \mathrm{c}\phi)\mathbf{u}\mathbf{u}' + \mathrm{s}\phi \, \tilde{\mathbf{u}}$$

Provided that dt is infinitesimal, ϕ is also infinitesimal and hence small angle approximations apply

$$\mathrm{c}\phi = \cos\phi = \cos(w\,dt) \cong 1$$
$$\mathrm{s}\phi = \sin\phi = \sin(w\,dt) \cong w\,dt$$

can be substituted in $\mathbf{C}_{b(t+dt)}^{b(t)}$ to get

$$\mathbf{C}_{b(t+dt)}^{b(t)} = \mathrm{c}\phi\mathbf{I} + (1 - \mathrm{c}\phi)\mathbf{u}\mathbf{u}' + \mathrm{s}\phi\,\tilde{\mathbf{u}} \cong \mathbf{I} + w\,dt\,\tilde{\mathbf{u}}$$

Equation (3.31) implies that

$$\mathbf{C}_b^a(t+dt) - \mathbf{C}_b^a(t) = \mathbf{C}_b^a(t)\left(\mathbf{C}_{b(t+dt)}^{b(t)} - \mathbf{I}\right)$$

Substituting for $\mathbf{C}_{b(t+dt)}^{b(t)}$, ignoring the higher order terms and dividing both sides by dt in the above equation gives

$$\frac{\mathbf{C}_b^a(t+dt) - \mathbf{C}_b^a(t)}{dt} = \mathbf{C}_b^a(t)\,w\tilde{\mathbf{u}} = \mathbf{C}_b^a(t)\,\tilde{\boldsymbol{\omega}}$$

Taking the limits of the left-hand side as dt approaches 0 gives

$$\frac{d\mathbf{C}_b^a(t)}{dt} = \mathbf{C}_b^a(t)\,\tilde{\boldsymbol{\omega}}$$

As a matter of formality the above equation may be written as

$$\dot{\mathbf{C}}_b^a = \mathbf{C}_b^a \tilde{\boldsymbol{\omega}}_{ab}^b \tag{3.34}$$

where the superscript b indicates that $\boldsymbol{\omega}$ is represented in frame $\boldsymbol{b}$, and the subscript ab indicates that the rotation vector is of frame $\boldsymbol{b}$ relative to frame $\boldsymbol{a}$.

To implement Eq. (3.34), given that a DCM is a 3×3 matrix, implies the need to solve nine differential equations. This raises two concerns: the computational load and the loss of the unitary property of the DCM. Nevertheless, there are computational methods to correct for the loss of its unitary, and with fast processors the computational load might not be of big concern.

Alternatives to the DCM differential equations are other attractive algorithms: the quaternions, the rotation vector and the Euler angles, which are all discussed next.

3.8.2 Quaternion Differential Equation

Here we follow precisely the steps taken in the previous derivation.

$$\mathbf{Q}_a^b(t)=\mathbf{Q}_a^{b(t)} \tag{3.35}$$

Thus at time $t+dt$

$$\mathbf{Q}_a^b(t+dt)=\mathbf{Q}_a^{b(t+dt)}=\mathbf{Q}_a^{b(t)}\mathbf{Q}_{b(t)}^{b(t+dt)}=\mathbf{Q}_a^b(t)\mathbf{Q}_{b(t)}^{b(t+dt)}$$

As before, we suppose that, in the time interval $[t,\ t+dt]$, the angular rate vector of frame $\boldsymbol{b}$ is

$$\boldsymbol{\omega}=w\,\mathbf{u} \tag{3.36}$$

where w is the angular rate and $\mathbf{u}$ is a unit vector represented in frame $\boldsymbol{b}$.

Then, in the time interval dt, frame $\boldsymbol{b}$ would rotate an angle given by

$$\phi=w\,dt \tag{3.37}$$

From the rotation properties we can write

$$\mathbf{Q}_{b(t)}^{b(t+dt)} = \left\{ c\frac{\phi}{2}, s\frac{\phi}{2}\,\mathbf{u} \right\} \tag{3.38}$$

where

$$c\,\frac{\phi}{2} = \cos(\frac{\phi}{2}) = \cos(\frac{w\,dt}{2}) \cong 1$$

and

$$s\frac{\phi}{2} = \sin(\frac{\phi}{2}) = \sin(\frac{w\,dt}{2}) \cong \frac{w\,dt}{2}$$

Substituting in the above equation yields

$$\mathbf{Q}_{b(t)}^{b(t+dt)} = \{c\frac{\phi}{2},\, s\frac{\phi}{2}\mathbf{u}\} \cong \{1, \frac{1}{2} w\mathbf{u}dt\} = \{1, \frac{1}{2}\boldsymbol{\omega}dt\}$$

$$\mathbf{Q}_a^b(t+dt) = \mathbf{Q}_a^b(t)\{1, \frac{1}{2}\boldsymbol{\omega}dt\} = \mathbf{Q}_a^b(t) + \mathbf{Q}_a^b(t)\ \{0, \frac{1}{2}\boldsymbol{\omega}dt\}$$

which implies that

$$\frac{\mathbf{Q}_a^b(t+dt)\text{-}\mathbf{Q}_a^b(t)}{dt} = \mathbf{Q}_a^b(t)\{0,\ \frac{1}{2}\boldsymbol{\omega}\}$$

Taking the limit of the left-hand side as dt approaches 0, gives

$$\frac{d\mathbf{Q}_a^b(t)}{dt} = \frac{1}{2}\mathbf{Q}_a^b(t)\,\{0, \boldsymbol{\omega}_{ab}^b\} \tag{3.39}$$

In matrix format the above equation becomes

$$\frac{d}{dt}\begin{bmatrix} q_o \\ \mathbf{q} \end{bmatrix} = \frac{1}{2}\begin{bmatrix} 0 & -\boldsymbol{\omega}' \\ \boldsymbol{\omega} & -\tilde{\boldsymbol{\omega}} \end{bmatrix}\begin{bmatrix} q_o \\ \mathbf{q} \end{bmatrix} \tag{3.40}$$

3.8.3 Rotation Vector Differential Equation

Like the quaternion, we derive a similar dynamic equation that relates the rotation vector to the angular rate vector, $\boldsymbol{\omega}$, [1,2]. This derivation is done indirectly through the quaternion equations.

First, we substitute for q_0 and $\mathbf{q}$ from Eq. (3.16) into Eq. (3.40) to get

$$\frac{d}{dt}\begin{bmatrix} q_o \\ \mathbf{q} \end{bmatrix} = \frac{1}{2}\begin{bmatrix} 0 & -\boldsymbol{\omega}' \\ \boldsymbol{\omega} & -\tilde{\boldsymbol{\omega}} \end{bmatrix}\begin{bmatrix} q_o \\ \mathbf{q} \end{bmatrix} = \frac{1}{2}\begin{bmatrix} 0 & -\boldsymbol{\omega}' \\ \boldsymbol{\omega} & -\tilde{\boldsymbol{\omega}} \end{bmatrix}\begin{bmatrix} \cos\left(\frac{\phi}{2}\right) \\ \sin\left(\frac{\phi}{2}\right)\mathbf{u} \end{bmatrix}$$

Performing the matrix product in the above yields

$$\frac{d}{dt}\begin{bmatrix} q_o \\ \mathbf{q} \end{bmatrix} = \frac{1}{2}\begin{bmatrix} -\sin\left(\frac{\phi}{2}\right)\boldsymbol{\omega}'\mathbf{u} \\ \cos\left(\frac{\phi}{2}\right)\boldsymbol{\omega} - \sin\left(\frac{\phi}{2}\right)\tilde{\boldsymbol{\omega}}\mathbf{u} \end{bmatrix} \tag{3.41}$$

Using Eq. (3.16) to get the time derivatives of q_0 and $\mathbf{q}$ gives

$$\begin{aligned} \dot{q}_O &= -\frac{\dot{\phi}}{2}\sin\left(\frac{\phi}{2}\right) \\ \dot{\mathbf{q}} &= \cos\left(\frac{\phi}{2}\right)\frac{\dot{\phi}}{2}\mathbf{u} + \sin\left(\frac{\phi}{2}\right)\dot{\mathbf{u}} \end{aligned} \tag{3.42}$$

Equating both sides of Eqs. (3.41) and (3.42) yields

$$\begin{aligned} -\frac{\dot{\phi}}{2}\sin\left(\frac{\phi}{2}\right) &= -\frac{1}{2}\sin\left(\frac{\phi}{2}\right)\boldsymbol{\omega}'\mathbf{u} \\ \cos\left(\frac{\phi}{2}\right)\frac{\dot{\phi}}{2}\mathbf{u} + \sin\left(\frac{\phi}{2}\right)\dot{\mathbf{u}} &= \frac{1}{2}\left(\cos\left(\frac{\phi}{2}\right)\boldsymbol{\omega} - \sin\left(\frac{\phi}{2}\right)\tilde{\boldsymbol{\omega}}\mathbf{u}\right) \end{aligned} \tag{3.43}$$

Simplifying and collecting terms, Eq. (3.43) becomes

$$\dot{\phi} = \boldsymbol{\omega}'\mathbf{u} \tag{3.44}$$

and

$$\sin\left(\frac{\phi}{2}\right)\dot{\mathbf{u}} = -\frac{1}{2}\sin\left(\frac{\phi}{2}\right)\tilde{\boldsymbol{\omega}}\mathbf{u} + \frac{1}{2}\cos\left(\frac{\phi}{2}\right)\left(\boldsymbol{\omega}\text{-}\dot{\phi}\mathbf{u}\right) \tag{3.45}$$

Substituting from Eq. (3.44) in the above and simplifying yields

$$\dot{\mathbf{u}} = -\frac{1}{2}\tilde{\boldsymbol{\omega}}\mathbf{u} + \frac{1}{2}\cot\left(\frac{\phi}{2}\right)\left[\boldsymbol{\omega}\text{-}(\boldsymbol{\omega}'\mathbf{u})\mathbf{u}\right] \tag{3.46}$$

Substituting for the cross vector identities

$$(\boldsymbol{\omega}'\mathbf{u})\mathbf{u} - \boldsymbol{\omega} = \mathbf{u}\times(\mathbf{u}\times\boldsymbol{\omega}) \tag{3.47}$$

and

$$-\tilde{\boldsymbol{\omega}}\mathbf{u} = -\boldsymbol{\omega}\times\mathbf{u} = \mathbf{u}\times\boldsymbol{\omega} \tag{3.48}$$

in Eq. (3.46) yields

$$\dot{\mathbf{u}} = \frac{1}{2}\mathbf{u}\times\boldsymbol{\omega} - \frac{1}{2}\cot\left(\frac{\phi}{2}\right)\mathbf{u}\times(\mathbf{u}\times\boldsymbol{\omega}) \tag{3.49}$$

Now, that we have the time derivatives of ϕ and $\mathbf{u}$ (the components of the rotation vector) it is now straightforward to determine the time derivative of the rotation vector $\overline{\boldsymbol{\varphi}} = \phi\mathbf{u}$. Evaluating the time derivative and substituting from Eqs. (3.44) and (3.49) yields

$$\begin{aligned}\dot{\overline{\boldsymbol{\varphi}}} &= \dot{\phi}\mathbf{u} + \phi\dot{\mathbf{u}} \\ &= (\boldsymbol{\omega}'\mathbf{u})\mathbf{u} + \frac{\phi}{2}\mathbf{u}\times\boldsymbol{\omega} - \frac{\phi}{2}\cot\left(\frac{\phi}{2}\right)\mathbf{u}\times(\mathbf{u}\times\boldsymbol{\omega})\end{aligned} \tag{3.50}$$

Substituting from the vector identity, $(\boldsymbol{\omega}'\mathbf{u})\mathbf{u} = \mathbf{u}\times(\mathbf{u}\times\boldsymbol{\omega}) + \boldsymbol{\omega}$, into Eq. (3.50) gives

$$\begin{aligned}\dot{\overline{\boldsymbol{\varphi}}} &= \mathbf{u}\times(\mathbf{u}\times\boldsymbol{\omega}) + \boldsymbol{\omega} + \frac{\phi}{2}\mathbf{u}\times\boldsymbol{\omega} - \frac{\phi}{2}\cot\left(\frac{\phi}{2}\right)\mathbf{u}\times(\mathbf{u}\times\boldsymbol{\omega}) \\ &= \boldsymbol{\omega} + \frac{\phi}{2}\mathbf{u}\times\boldsymbol{\omega} + \left(1 - \frac{\phi}{2}\cot\left(\frac{\phi}{2}\right)\right)\mathbf{u}\times(\mathbf{u}\times\boldsymbol{\omega})\end{aligned}$$

Substituting for $\mathbf{u}$ from Eq. (3.6) gives

$$\dot{\overline{\boldsymbol{\varphi}}} = \boldsymbol{\omega} + \frac{1}{2}\overline{\boldsymbol{\varphi}}\times\boldsymbol{\omega} + \frac{1}{\phi^2}\left(1 - \frac{\phi}{2}\cot\left(\frac{\phi}{2}\right)\right)\overline{\boldsymbol{\varphi}}\times(\overline{\boldsymbol{\varphi}}\times\boldsymbol{\omega}) \tag{3.51}$$

In situations where ϕ is small, the third term in Eq. (3.51) becomes negligible. When omitted, the rotation vector equation becomes

$$\dot{\bar{\boldsymbol{\varphi}}} = \boldsymbol{\omega} + \frac{1}{2}\bar{\boldsymbol{\varphi}} \times \boldsymbol{\omega} \tag{3.52}$$

From a physical standpoint, consider a body that goes through a continuous rotation and consider its rotation in the infinitesimal time interval, t to $t+dt$. The body rotation $\bar{\boldsymbol{\varphi}}$ in this interval (the body rotation at $t+dt$ relative to its attitude at time t) is infinitesimal and hence Eq. (3.52) becomes

$$\dot{\bar{\boldsymbol{\varphi}}} = \boldsymbol{\omega}$$

Therefore the attitude rate is the same as the angular rate vector, $\boldsymbol{\omega}$, and hence $\boldsymbol{\omega}$ is called the 'instantaneous rotation vector'. The above equation is consistent with Equations (3.33) and (3.37) that were obtained for the DCM and the quaternion, respectively.

3.8.4 Euler Angles Differential Equation

The Euler angles derivatives are obtained as functions of the body angular rate with respect to some reference frame as follows. We first observe that the Euler angles, and consequently their derivatives, are performed in different frames (see Fig 3.1). If we reference all of them to the body frame, we get

$$\boldsymbol{\omega} = \mathbf{C}_{x''}(\phi)\begin{bmatrix}\dot{\phi}\\0\\0\end{bmatrix} + \mathbf{C}_{x''}(\phi)\mathbf{C}_{y'}(\theta)\begin{bmatrix}0\\\dot{\theta}\\0\end{bmatrix} + \mathbf{C}_{x''}(\phi)\mathbf{C}_{y'}(\theta)\,\mathbf{C}_{z}(\psi)\begin{bmatrix}0\\0\\\dot{\psi}\end{bmatrix} \tag{3.53}$$

Expanding and simplifying yields

$$\boldsymbol{\omega}=\begin{bmatrix}\omega_x\\ \omega_y\\ \omega_z\end{bmatrix}=\begin{bmatrix}\dot{\phi}\text{ - }\dot{\psi}\,\mathrm{s}(\theta)\\ \dot{\theta}\,\mathrm{c}(\phi)+\dot{\psi}\,\mathrm{s}(\phi)\,\mathrm{c}(\theta)\\ -\dot{\theta}\,\mathrm{s}(\phi)+\dot{\psi}\,\mathrm{c}(\phi)\,\mathrm{c}(\theta)\end{bmatrix} \tag{3.54}$$

Solving for the Euler angles derivatives from Eq. (3.54) gives

$$\begin{bmatrix}\dot{\phi}\\ \dot{\theta}\\ \dot{\psi}\end{bmatrix}=\begin{bmatrix}\omega_x+(\omega_y\,\mathrm{s}\phi+\omega_z\,\mathrm{c}\phi)\tan\theta\\ \omega_y\,\mathrm{c}\phi-\omega_z\,\mathrm{s}\phi\\ (\omega_y\,\mathrm{s}\phi+\omega_z\,\mathrm{c}\phi)/\mathrm{c}\theta\end{bmatrix} \tag{3.55}$$

From the above, it is evident that the Euler angle approach is attractive since it requires only three elements to integrate as compared with nine elements in the DCM and four elements in the quaternion. Since there is no redundancy, the unitary property of the transformation matrix is naturally preserved. Nevertheless it suffers serious drawbacks that make it less appealing. The derivatives in Eq. (3.55) use trigonometric functions heavily, a very time consuming process. More importantly, the division by cos(θ) results in very large derivatives when θ approaches 90 degrees.

We have introduced the different forms for vector transformation and derived their respective dynamic equations. We also provided the equations that convert between these transformations. The strong similarity between the quaternion and the rotation vector has been demonstrated. With this analytical background we are almost ready to tackle the derivation of the inertial navigation equation. Before we do that we must do a little analysis of the surface over which our vehicles shall navigate: the Earth. The surface of the Earth will be approximated with a special geometric shape (the ellipsoid) to which the position coordinates will be referenced. This is the subject of the following chapter.

References

1. Bortz, J.E.," A New Mathematical Formulation for Strapdown Inertial Navigation," IEEE Transactions on Aerospace and Electronic Systems, Vol. AES-7, No. 1, 1971, pp. 61-66.
2. Y. F. Jiang and Y. P. Lin, "On the Rotation Vector Differential Equation," IEEE Transactions on Aerospace and Electronic Systems, Vol. AES-27, No. 1, 1991, pp. 181-183.

Chapter 4

Earth and Navigation

4.1 Introduction

The history of navigation is intertwined with Earth. Not only because navigation is associated with the Earth's shape (spherity, landmarks), but also because it is influenced by the Earth's dynamics (gravity, rotation). People navigating the Earth from the beginning of mankind found many ways and forms though none as sophisticated as our modern means of navigation which entail such techniques as stellar, radio, GPS and inertial navigation. Our attention is entirely focused on the latter: inertial navigation.

Like all methods, inertial navigation has its advantages and shortcomings. Inconsistent accuracy — resulting from its time varying errors — is one of its greatest disadvantages. Nonetheless, what makes it unique and attractive is its self-reliance: its function is independent on outside sources. Short of power failure or malfunction of one of its sensors, this system can work indefinitely. At its core are a set of accelerometers, a set of gyros and a computer. The computer, by means of the gyros, tracks the craft's attitude, as in general it does not remain straight and level. The computer also integrates the accelerometer outputs to compute the velocity and once more to compute the position.

But this is a bit oversimplified. First, the accelerometers operate in a Newtonian Gravitational field and as such their measurements of the craft accelerations are biased with the Earth's gravity. Therefore, the sensor readings must be corrected for the gravity bias to yield meaningful results. Second, the gyros are also inertial sensors that measure the Earth's rotation on top of the craft's attitude. Thus, the Earth's rotation must be tracked and accounted for at the gyro outputs. In addition, gyros also sense an extra component. Because of the curvature of the Earth, the craft's motion is actually rotational about the surface of the Earth. This rotation is also sensed by the gyros and must also be accounted for when computing the craft's attitude. Finally, navigating crafts invariably adhere to the surface of the Earth or fly closely to it. This surface is neither flat nor smooth hence a reference system for shaping the Earth is in order.

The above merely points to the need of quantifying Earth's geometry, its rotational rate and gravity. With these elements specified, we should be able to define a universal reference system to which we can determine a craft's position and velocity at any point on Earth. In the following sections we will describe the Earth geometry and its two important elements, the gravity and rotational rate. The history of the Earth's geometry is as wide and deep as that of the human race. Therefore, we will address only and very briefly its current state and the elements that affect our topic.

4.2 Earth, Geoid and Ellipsoid

Geodesists have associated three models to Earth: topographical, geodetic, and ellipsoidal. The topographical model is the actual face of the Earth, the one that we physically see in our daily life. The model is not smooth and not associated with any geometrical shape. For navigation purposes it can be extremely difficult to work with to determine one's location and hence is not discussed further. What we need is a smooth geometrical surface with the gravity defined at each point. Ultimately, this smooth surface is an ellipsoid that closely matches the face of the Earth and to which the position coordinates of any object will be referenced. Before we discuss the ellipsoid, we need to address two central concepts on which its geometry is based: the mean sea level (MSL) and the equipotential surfaces. The MSL is the water level that would cover Earth in absence of wave tides and wind; it will define the face of the Earth we have just alluded to.

In the sense of assigning locations, what would surveyors do to survey objects at a certain site? They select a point, o, to serve as a reference or datum to which all points will refer. This point is associated with a "level" to which the height of other points can be measured. As such, all points of zero height will be at the same level as o. For example (adopting a North, East, Down frame) a point with coordinates (x, y, z) would mean it is x meters north of o, y meters east of o and of height z meters below o.

But what do "level" and "height" mean? On the Earth's surface, leveling is a central concept because all surrounding marks will be assigned a height according to their levels. On large mass areas, the mean sea level plays the role of the reference level. However, in actual surveying, a level surface to geodesists is that on which gravity is normal to it, or in other words, has no tangential gravity components. This surface is called equipotential, a surface of constant gravity potential. Any point (on, above or below Earth's surface) lies on an equipotential surface. That is, there is a continuum of three dimensional contours each assigned a specific potential value. Because of this property, these surfaces are of interest because they essentially define what is "up" and "down" and consequently height and elevation. The equipotential contour that "best fits" the MSL in the least square sense is called the geoid and has replaced the MSL as a reference for leveling (see Fig. 4.1) [1]. Elevation is reserved to indicate height normal to the geoid while height indicates the distance normal to the ellipsoid. It should be noted that the geoid is not a hypothetical or derived surface, rather a surface associated with physical data.

Have we solved the height of objects at any point on or close to Earth? Not quite. Geoid is a smooth surface but its shape is irregular and does not provide the smooth geometrical shape needed for navigation. So what geodesists did is fit the geoid to an ellipsoid with very special properties. This ellipsoid (also called the normal ellipsoid) has an equipotential surface, has the same mass and angular velocity as that of Earth and whose geometric center coincides with the center of mass of Earth [2]. The relation between the Earth, geoid and ellipsoid surfaces is depicted in Fig. 4.2. Variations in height from the geoid to the ellipsoid are called geoid undulations.

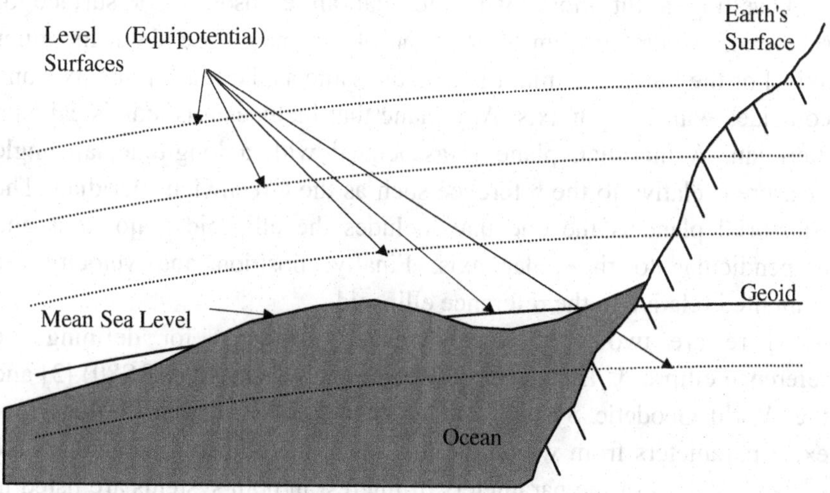

Figure 4.1 Mean Sea Level and Equipotential Levels

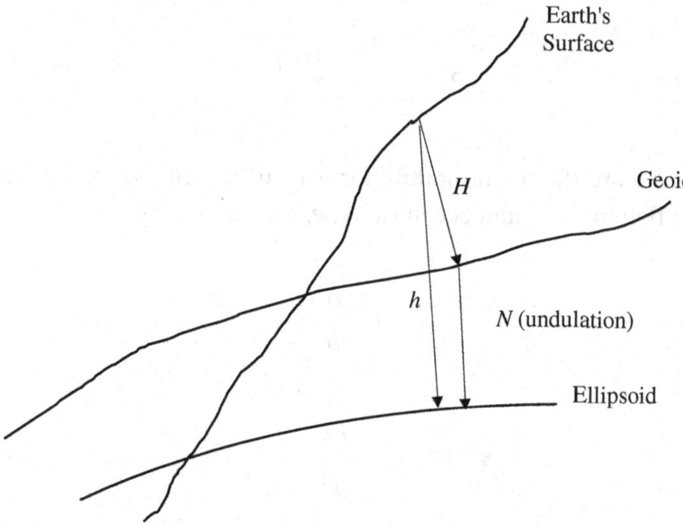

Figure 4.2 Relative Heights between Earth, Geoid and Ellipsoid

Dwelling a bit more, we note that an ellipsoid is a surface of revolution generated from the rotation of a plane ellipse about its minor axis. For the Earth, the minor axis is the same as the Earth polar axis and coincides with its spin axis. Any plane that includes this axis is called a meridian. A meridian plane is associated with a longitude, an angle measured relative to the reference such as the Greenwich Meridian. The equatorial plane is the one that includes the ellipsoid major axis and perpendicular to the polar axis. Finally, position and velocity are computed relative to the reference ellipsoid.

There are two very closely related standards for defining the reference ellipsoid. The Geodetic Reference System 80 (GRS80) [3] and the World Geodetic System (WGS84) [4]. Each standard defines four exact parameters from which the rest of all the ellipsoid parameters are derived. Values of the parameters of interest in both systems are listed in Table 4.1. We briefly list some of the ellipse and ellipsoid geometric properties, which are described in more detail in Appendix D. First, we note that an ellipse is described by the equation

$$\frac{x^2}{a^2} + \frac{y^2}{b^2} = 1 \tag{4.1}$$

where a and b are the major and minor axes of the ellipse, respectively. The ellipse flattening, f, and eccentricity, e, are defined by

$$f = 1 - \frac{b}{a} \tag{4.2}$$

$$e^2 = 1 - \frac{b^2}{a^2} \tag{4.3}$$

Table 4.1

GRS80 and WGS84 Parameters

Parameter	Unit	GRS80	WGS84
a, semi-major axis	m	6 378 178 (exact)	6 378 178 (exact)
Ω_e, Angular velocity	10^{-11} rad s^{-1}	7 292 115(exact)	7 292 115 (exact)
GM, Geocentric gravitational const.	10^{-8} m^3s^{-2}	3 986 005(exact)	3 986 005 (exact)
1/*f*, Reciprocal flattening	-	298.257 222 101	298.257 223 563(exact)
b, Semi-minor axis	m	6 356 752.3141	6 356 752.3142
$m = \frac{\Omega_e^2 a^2 b}{GM}$, Dimensionless const.	-	.003 449 786 003 08	.003 449 786 506 84
γ_e, normal gravity at equator	ms^{-2}	9.780 326 7715	9.780 325 3359
γ_p, normal gravity at pole	ms^{-2}	9.832 186 3685	9.832 184 9378
e2, first eccentricity(e)	-	.006 694 380 022 90	.006 694 379 990 14
J2, dynamic form factor	-	108 263x10^{-8}	

4.3 Radii of Curvature

A craft that moves on the surface of Earth, ideally the reference ellipsoid, is actually rotating about it. Determining the radii of curvature is essential for computing the craft rotation and consequently its position. Because the Earth is not a perfect sphere, then at any point on its surface there are two radii of curvature. The first is sensed when a craft moves north-south along a meridian, and the other is sensed when a craft moves east-west. As will be seen shortly, these radii are functions of the geodetic latitude. As shown in Fig. 4.3 there are two latitudes to be distinguished. The first is the geodetic latitude, ϕ, which is the angle between the vertical line at *c* (plump bob) and the equatorial plane. The other is the geocentric latitude, ϕ_c, which is the angle between the line *oc* (that joins the point *c* and *o*, the Earth's center) and the equatorial plane. The geodetic latitude shall be the one used in our computations.

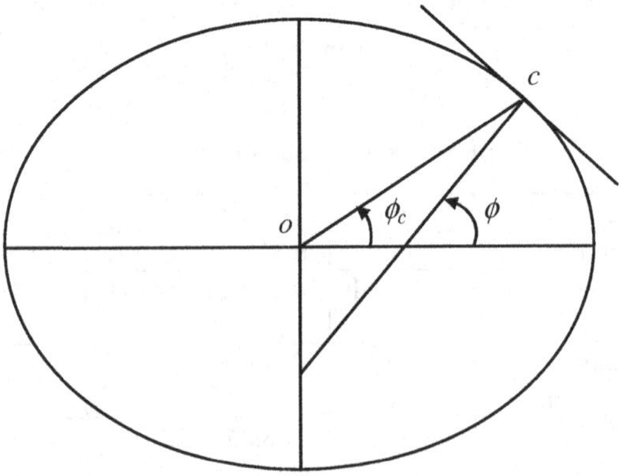

Figure 4.3 Geocentric and Geodetic Latitudes

We now discuss the radii of curvatures at some point on the Ellipsoid. The meridian radius of curvature, R_m, (see Fig. 4.4a) is the radius of curvature in the meridian plane. The prime radius of curvature, R_p, (see Fig. 4.4b) is the radius of curvature in the plane normal to the meridian (i.e. the plane in which the craft moves east-west).

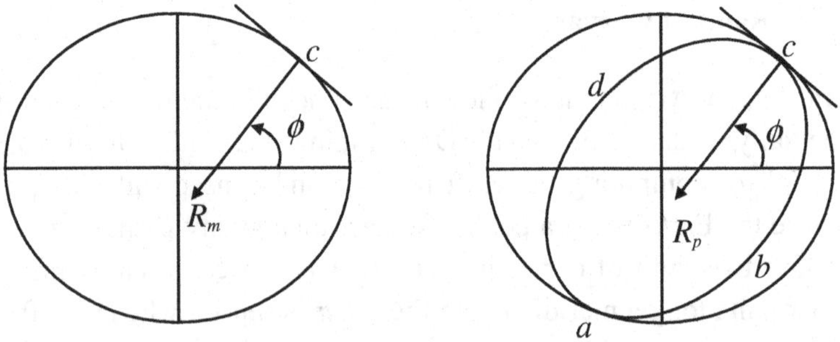

a. Meridian radius of curvature, R_m, in the meridian plane

b. Prime radius of curvature, R_p, in the ellipse 'abcd', normal to the meridian

Figure 4.4 Earth Radii of Curvature

For a point on the Ellipsoid at which the geodetic latitude is ϕ (see Appendix D) R_m, R_p are given by

$$R_m = \frac{a(1-e^2)}{\left[1-e^2\sin^2\phi\right]^{3/2}} \tag{4.4}$$

$$R_p = \frac{a}{\left[1-e^2\sin^2\phi\right]^{1/2}} \tag{4.5}$$

4.4 Earth, Inertial and Navigation Frames

Why do we need all these frames? Mainly, to simplify computations and render them relevant to our need. We briefly discuss the purposes of these frames (where they arise) and then define them afterwards. The inertial frame is the one in which all inertial measurements are referenced to. The Earth frame is the one in which the Earth spins about itself and is generally associated with the normal ellipsoid. The craft position relative to the Earth frame can be computed, but it will be extremely inconvenient to use for real time applications. The navigation frame is more convenient, as it determines the position in terms of longitude and latitude. These frames are defined as follows:

1. The inertial frame, for simplicity, (see Fig. 4.5) is a stationary frame centered to Earth and shares its polar axis but does not rotate with it. The inertial frame plays a central role in navigation because inertial sensors measure the craft motion relative to it. From a practical standpoint, inertial motion will be confined to Earth's rotation, body motion about the spherical Earth, and body rotation about itself.

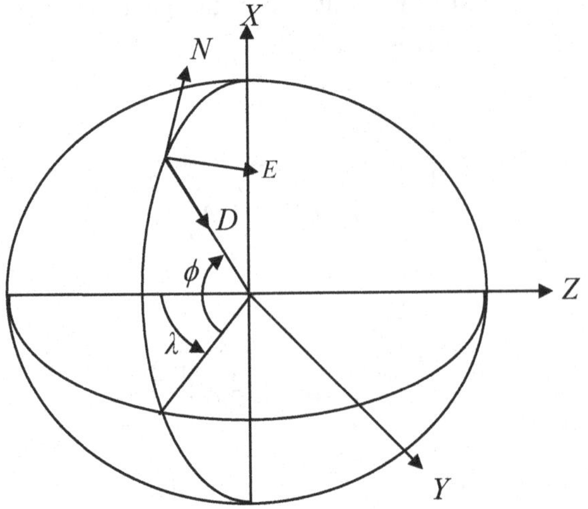

Figure 4.5 Earth and Navigation Frames

2. The Earth frame is Earth-centered, designated with *XYZ*, is one in which the *XZ* plane is in the reference meridian plane, with the *X*-axis pointing to the North Pole. The *Z*-axis is at the intersection of the Greenwich meridian and the equatorial plane, pointing away from the center. Finally, the *Y*-axis is in the equatorial plane and perpendicular to the reference meridian plane. Although this is not the standard WGS84 definition for the Earth frame, it is adopted here to conform to the North-East-Down (NED) axes system. We obtain the NED frame by merely two Euler rotations: the first is to rotate the Earth frame an angle λ about the *X*-axis and next to rotate it by an angle $-\phi$ about the *Y*-axis.

3. The navigation frame (Nav-frame) , designated with *xyz*, is one that moves with the craft about the surface of the Earth, which in reality is the ellipsoid. The *z*-axis is normal to the ellipsoid at the location of the craft, and points down to the inside of the ellipsoid. The *xy* plane is tangent to the ellipsoid at the intersection of the *z*-axis and the ellipsoid, with the *x* and *y* axes pointing north and east, respectively. The *z*-axis intersects the equatorial plane at the geodetic latitude angle, ϕ. The transformation matrix from the Earth to Nav-frame can be given by

$$\mathbf{C}_e^n = \mathbf{C}_y(-\phi)\,\mathbf{C}_x(\lambda)$$

$$= \begin{bmatrix} \cos\phi & -\sin\phi\sin\lambda & \sin\phi\cos\lambda \\ 0 & \cos\lambda & \sin\lambda \\ -\sin\phi & -\cos\phi\sin\lambda & \cos\phi\cos\lambda \end{bmatrix} \tag{4.6}$$

Incidentally, we could have selected this Nav-frame to be fixed to the Earth and pointed North-East-Down at the original point of departure. This would be perfectly fine for short-term navigation in which Earth sphericity is not an issue. But for long-term navigation, the position and velocity computed in this frame will not be of practical use without further processing. The differential equation of this DCM is given in the following section.

4.5 Earth Rate

The Earth rotation rate spins about its polar axis at a rate of Ω_e rad/sec (given in Table 4.1). This rotation represents the Earth's angular rate relative to the inertial space, and in Earth frame it is represented by $\boldsymbol{\omega}_{ie}^e = (\Omega_e \quad 0 \quad 0)'$. Using the transformation matrix $\mathbf{C}_e^n$ given by Eq. (4.6), this rate can be expressed in the Nav-frame by

$$\boldsymbol{\omega}_{ie}^n = \mathbf{C}_e^n \boldsymbol{\omega}_{ie}^e$$

$$= \begin{bmatrix} \cos\phi \\ 0 \\ \text{-}\sin\phi \end{bmatrix} \Omega_e \tag{4.7}$$

4.6 The Craft Rate $\boldsymbol{\omega}_{en}^n$

To determine the craft's position on a spherical surface it is easier and more convenient to use angular coordinates than Cartesian coordinates.

For example, consider a particle that moves on the perimeter of a unit circle with two perpendicular axes at its center. One would only need its angular coordinate relative to one of these axes to determine its position instead of two Cartesian coordinates. However, to apply the spherical coordinates, we must convert the craft's linear velocities to angular rates. Consider a craft that moves on the surface of the ellipsoid (zero height) and whose velocity components along the north and east directions are v_n and v_e, respectively. In this case, the craft's angular rate about the east axis (due to v_n) is $\dot{\phi} = \frac{v_n}{R_m}$ where R_m (given by Eq. (4.4)) is the meridian radius of curvature (see Fig. 4.6a). Likewise the craft's angular rate about the polar axis (due to v_e) is $\dot{\lambda} = \frac{v_e}{R_p \cos\phi}$ where R_p (given by Eq. (4.5)) is the prime radius of curvature in the plane normal to the meridian (see Fig 4.6b). At height, h, above the ellipsoid, the craft's angular rates are

$$\dot{\phi} = \frac{v_n}{R_m + h} \tag{4.8}$$

and

$$\dot{\lambda} = \frac{v_e}{(R_p + h)\cos\phi} \tag{4.9}$$

To express these angular rates in the Nav frame, the angular rate about the polar axis, $\dot{\lambda}$, should be resolved along the north and down axes. The angular rate vector, $\boldsymbol{\omega}_{en}^{n}$, of the Nav-frame about the Earth frame becomes

$$\boldsymbol{\omega}_{en}^{n} = \begin{bmatrix} \dot{\lambda}\cos\phi \\ -\dot{\phi} \\ -\dot{\lambda}\sin\phi \end{bmatrix} \tag{4.10}$$

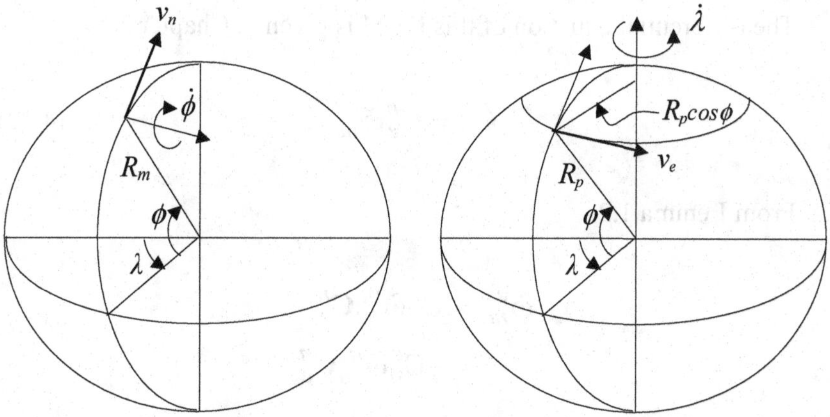

a. North-South motion in the meridian plane

b. East-West motion in a plane parallel to the equatorial plane

Figure 4.6 Craft's angular rates

Substituting for $\dot{\lambda}$ and $\dot{\phi}$ in the above equation, we get the craft rate equation

$$\boldsymbol{\omega}_{en}^{n} = \begin{bmatrix} \dfrac{v_e}{R_p + h} \\ -\dfrac{v_n}{R_m + h} \\ -\dfrac{v_e \tan\phi}{R_p + h} \end{bmatrix} \tag{4.11}$$

4.7 Solution of the DCM $\mathbf{C}_e^n$

Computing $\mathbf{C}_e^n$ is central to the navigation algorithm because it embeds the information that determines the craft's longitude and latitude. Further it is needed to solve for the Earth rate and the craft's rate given in the previous two sections.

The differential equation of this DCM is given in Chapter 3 by

$$\dot{\mathbf{C}}_e^n = \mathbf{C}_e^n \tilde{\boldsymbol{\omega}}_{ne}^e \tag{4.12}$$

From Lemma 1.2

$$\begin{aligned}\tilde{\boldsymbol{\omega}}_{ne}^e &= \mathbf{C}_n^e \tilde{\boldsymbol{\omega}}_{ne}^n \mathbf{C}_e^n \\ &= -\mathbf{C}_n^e \tilde{\boldsymbol{\omega}}_{en}^n \mathbf{C}_e^n\end{aligned} \tag{4.13}$$

Substituting for the above in Eq. (4.12), we obtain

$$\dot{\mathbf{C}}_e^n = -\tilde{\boldsymbol{\omega}}_{en}^n \mathbf{C}_e^n \tag{4.14}$$

4.8 Gravitational and Gravity Fields

It will be evident in the next chapter that in order to implement the navigation equations, a mathematical model for the gravity vector is needed. Before we discuss this model we will introduce the two components of the gravity vector: the gravitational field and the centrifugal force.

Gravitational field vector, $\mathbf{g}_m$, is the force of attraction between the Earth and a unit mass outside its surface. This is the only force (ignoring attractions by all other heavenly bodies) that would be exerted on a unit mass that is at rest in the inertial space (not rotating with Earth). If the mass is rotating along with the Earth then it will be additionally influenced by the centrifugal force (that results from the rotation about the Earth's polar axis). It can be seen that the centrifugal force is zero at the poles and a maximum in the equatorial plane. The gravity field vector, **g**, is the sum of the gravitational field and the centrifugal force

$$\mathbf{g} = \mathbf{g}_m - \boldsymbol{\omega}_{ie} \times \boldsymbol{\omega}_{ie} \times \mathbf{r} \tag{4.15}$$

where **r** is the geocentric position vector of the mass.

What we need is to relate the gravity vector (magnitude and direction) to the normal ellipsoid that we adopted as a reference for navigation. We first note that the equipotential surfaces, in particular the geoid, are not in general parallel to the surface of the ellipsoid. As such, the gravity vector being normal to the geoid, will be slightly off the normal to the ellipsoid, and in the NED frame it will be represented by

$$\mathbf{g} = \begin{bmatrix} \xi g \\ -\eta g \\ g \end{bmatrix} \tag{4.16}$$

where g is the magnitude of the gravity normal to the ellipsoid and ξ and η are the two angular displacements of the gravity vector from the normal to the ellipsoid.

One fundamental property of the normal ellipsoid is that the magnitude of the gravity normal to its surface (called normal gravity) is given by the closed form Somigliana formula [5]. The normal gravity, γ, at a point whose geodetic latitude, ϕ, is given by the

$$\gamma = \gamma_e \frac{1 + k \sin^2 \phi}{\sqrt{1 - e^2 \sin^2 \phi}} \tag{4.17}$$

where

$$k = \frac{b\,\gamma_p}{a\,\gamma_e} - 1 \qquad (4.18)$$

and γ_e and γ_p are the Earth's normal gravity magnitudes at the equator and the pole, respectively. The mathematical gravity model can then be described by

$$\boldsymbol{\gamma} = \begin{bmatrix} 0 \\ 0 \\ \gamma \end{bmatrix} \qquad (4.19)$$

The relationship between the actual gravity, **g**, and the model counterpart, **γ**, is depicted in Fig. 4.7. We see that there is a difference in magnitude and direction between the gravity vector, **g**, and the mathematical model given by

$$\Delta\mathbf{g} = \mathbf{g} - \boldsymbol{\gamma} = \begin{bmatrix} \xi g \\ -\eta g \\ \Delta g \end{bmatrix} \qquad (4.20)$$

where $\Delta g = g - \gamma$. The deviation in magnitude, Δg, is called the gravity anomaly, while the deviation of the gravity vector direction from the normal to the ellipsoid is called deflection of the vertical.

At points above the surface of the ellipsoid, the normal gravity vector will be slightly off the normal to the ellipsoid and have an additional component along the north direction. For a point h meters above the ellipsoid, the downward component of the normal gravity is given by

$$\gamma_h = \gamma\left[1 - \frac{2}{a}(1 + f + m - 2f\sin^2\phi)h\right] \qquad (4.21)$$

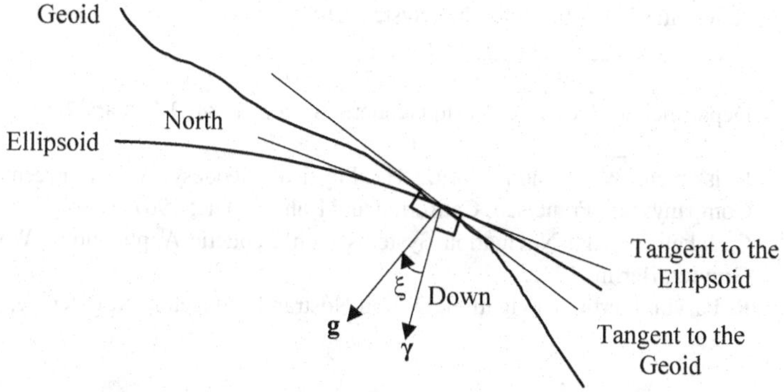

Figure 4.7 Relationship between **g** and γ in the North-Down plane

where the parameters *a, b, e, f, m* are as defined in Table 4.1. The normal gravity north component is given by [5,6]

$$\gamma_n(\phi) \approx -8.08\times10^{-6}\, h_{km} \sin(2\phi) \quad \mathrm{ms}^{-2} \tag{4.22}$$

where h_{km} is the height of the point in km units. In vector form, the normal gravity is given by

$$\boldsymbol{\gamma}_h = \begin{bmatrix} \gamma_n \\ 0 \\ \gamma_h \end{bmatrix} \tag{4.23}$$

An alternative to Eq. (4.22) is given in terms of angular deflection [7]

$$\delta_g = 0.0001585\, h \sin(2\phi) \quad arc\,\mathrm{sec}$$

References

1. What is the geoid? National Geodetic Survey
http://www.ngs.noaa.gov/GEOID/geoid_def.html
2. Xiong Li and Hans-Jurgen Gotze, 'Tutorial Ellipsoid, geoid, gravity and geophysics' Geophysics, Vol_66, No. 6 (November-December 2001) p. 1660-1668
http://www.lct.com/technical-pages/pdf/Li_G_Tut.pdf

3. H. Moritz, Geodetic Reference System 1980,
http://www.gfy.ku.dk/~iag/handbook/geodeti.htm
http://www.fgg.uni-lj.si/~/mkuhar/zalozba/GRS80_Moritz.pdf
4. Department of Defense, World Geodetic System 1984, 3 January 2000,
NGA: DoD World Geodetic System 1984
5. Heiskanen, W. A. and Moritz, H.; Physical Geodesy; W. H. Freeman and Company; San Francisco, California and London, UK; 1967.
6. C. Jekeli, Inertial Navigation Systems with Geodetic Applications, Walter de Gruyter, Berlin, 2001.
7. R. Parvin, Inertial Navigation, D. Van Nostrand, Princeton, New Jersey, 1962.

Chapter 5

The Inertial Navigation System Equations

5.1 Introduction

Navigation is simply the science or more accurately the art of computing the location and the velocity of a craft on land, sea or space. Depending on its mission, the craft is equipped with certain sensors to achieve such a task. For example, a Global Positioning System (GPS) receiver may be enough for a land or sea cruisers, but an aircraft will require far more instrumentation, namely air pressure sensors to measure altitude and airspeed. Although magnetic sensors are still used to find azimuth, modern aircraft now widely use inertial navigation system.

Inertial navigation systems are very popular because of their accuracy, long mean time between failure (MTBF), and self-reliance. Short of power supply they do not rely on any additional external sources for their operation. These systems use inertial sensors – accelerometers and gyros – to detect linear and angular motion of the craft. Incidentally, inertial sensors acquired this name because the motion they sense is with respect to a universal non-moving (inertial) space.

The mechanical structure of the original inertial systems was very elaborate and very sophisticated. In such a structure, the inertial sensors were mounted on a stable platform, which itself was mounted in the center of three (or more) perpendicular gimbals. These gimbals were used to allow the stable platform to maintain a certain attitude in the space. Although these systems were very accurate they also were highly expensive to manufacture and maintain. Soon, as computer technology matured, the much cheaper strapped down systems became more favorable. As its name might suggest, a strapped down system has no gimbals. In contrast to a gimbaled system, the inertial suite of a strapped down system is mechanically fixed to the craft. The inertial suite comprises two triads of mutually orthogonal inertial sensors, one for the accelerometers and the other for the gyros.

In the following we shall derive the navigation equations that, with the inputs of the inertial sensors, enables us to compute our position and velocity in the three dimensional space. Before we delve into deriving our equations, we must define one more frame of reference: the body frame.

5.2 Body Frame of Reference

In the previous chapter we defined:

i. the inertial frame to be Earth centered but not rotating with its axis. This frame is needed to identify the inertial space in which all bodies move relative to it and all the inertial sensor measurements are made relative to it.
ii. the Earth frame is centered to Earth and fixed to it and,
iii. the Nav-frame, which tracks the craft's motion on surface of the Earth.

We now define the body frame as a frame that is fixed to the craft body. For an aircraft, the x-axis points forward along its symmetric axis, the y-axis is orthogonal to the x-axis and points along the right wing, and the z-axis is orthogonal the xy plane and points downwards. When the craft flies straight and level with its x-axis pointing towards the north, this body frame will coincide with the North, East, and Down (NED) Nav-frame. Further, its z-axis would point along the normal gravity vector. With this arrangement, the accelerometer measurement would only be that due to gravity.

The transformation matrix from the Nav-frame to the body frame is constructed by three rotational Euler angles ψ, θ and ϕ about the z, y' and x'' axes respectively (see Fig. 3.1 in Section 3.2 for description and definition). Therefore

$$\mathbf{C}_n^b = \mathbf{C}_{x''}(\phi)\mathbf{C}_{y'}(\theta)\mathbf{C}_z(\psi) \tag{5.1}$$

5.3 Inertial Sensors

Accelerometers and gyros are members of the inertial suite that work hand in hand to make inertial navigation possible. Primarily, the functions of the accelerometer and gyro are sensing the craft's acceleration and its attitude, respectively. As mentioned earlier, inertial sensors sense the craft motion relative to the inertial space. Thus, they not only sense the body acceleration and rotation relative to Earth but also the Earth's gravity and rotation rate.

5.3.1 The Accelerometer

The accelerometer is a specific force sensor that senses both the craft inertial acceleration $\mathbf{a}_i$, as well as the gravitational field vector $\mathbf{g}_m$ (the force of mass attraction to the Earth). Therefore the accelerometer sensed craft specific force **a**, is given by

$$\mathbf{a} = \mathbf{a}_i - \mathbf{g}_m \tag{5.2}$$

If the particle position vector in the inertial space is **r**, then the inertial acceleration $\mathbf{a}_i$ is the second derivative of **r** with respect to time

$$\mathbf{a}_i = \left.\frac{d^2\mathbf{r}}{dt^2}\right|_i \tag{5.3}$$

where in the above equation, the subscript i is interpreted as "observed in the i-frame". Equation (4.15) shows that the gravitational field vector, the gravity field vector **g**, and the centripetal acceleration are related by

$$\mathbf{g}_m = \mathbf{g} + \boldsymbol{\omega}_{ie} \times \boldsymbol{\omega}_{ie} \times \mathbf{r} \tag{5.4}$$

where $\boldsymbol{\omega}_{ie}$ is the Earth angular velocity. Substituting from Eqs. (5.3) and (5.4) into (5.2) yields

$$\mathbf{a} = \left.\frac{d^2\mathbf{r}}{dt^2}\right|_i - \mathbf{g} - \boldsymbol{\omega}_{ie} \times \boldsymbol{\omega}_{ie} \times \mathbf{r} \tag{5.5}$$

5.3.2 The Rate Gyro

Rate gyros are sensors that measure angular velocities in contrast to attitude angles measured by free gyros, which are typically mounted in gimbaled platforms. Rate gyros sense the craft rate relative to the inertial space. That is they sense all these components: the craft angular rate relative to the Earth $\boldsymbol{\omega}_{nb}$, its angular rate as it moves about the spherical Earth $\boldsymbol{\omega}_{en}$ and the angular rate of the Earth itself $\boldsymbol{\omega}_{ie}$. The vector sum of these angular rates $\boldsymbol{\omega}_{ib}$ is given by

$$\boldsymbol{\omega}_{ib}^{b} = \boldsymbol{\omega}_{ie}^{b} + \boldsymbol{\omega}_{en}^{b} + \boldsymbol{\omega}_{nb}^{b} \tag{5.6}$$

For vector sum consistency, the above components must be represented in the same frame, regardless of which frame. The frame of choice is that indicated by the superscript. For our purposes we select the body frame, since it is the one in which the inertial rates are measured.

5.4 The Attitude Equation

Recall that a strapped down system is one in which the inertial sensors are rigidly mounted to the body of the aircraft. Thus, in particular, sensed accelerations are measured in the body frame. These accelerations will not be helpful to navigation unless they are transformed to the navigation frame. And that is where the rate gyros will be needed; they will serve as the mechanism for generating the needed transformation. Here we derive the equation that computes the body to navigation transformation.

Now how can the craft attitude be determined from these rate measurements? One might argue that $\boldsymbol{\omega}_{ie}$ the Earth rotation rate is well known. Also the craft rate $\boldsymbol{\omega}_{en}$– which is roughly the ratio of the craft velocity to the Earth radius – can be computed by dividing the integral of the craft acceleration by the Earth radius. This leaves us with $\boldsymbol{\omega}_{nb}$. But the variables $\boldsymbol{\omega}_{ie}$ and $\boldsymbol{\omega}_{en}$ in Eq. (5.6) are needed to be in the body frame while they can be computed only in the navigation frame. Thus we must determine the transformation $\mathbf{C}_b^n$ to convert these two variables from the navigation frame into the body frame. Using Eq. (3.32) one can write

$$\dot{\mathbf{C}}_b^n = \mathbf{C}_b^n \tilde{\boldsymbol{\omega}}_{nb}^b$$

Substituting for $\boldsymbol{\omega}_{nb}^b = \boldsymbol{\omega}_{ib}^b - \boldsymbol{\omega}_{in}^b$ in the above gives

$$\begin{aligned}\dot{\mathbf{C}}_b^n &= \mathbf{C}_b^n (\tilde{\boldsymbol{\omega}}_{ib}^b - \tilde{\boldsymbol{\omega}}_{in}^b) \\ &= \mathbf{C}_b^n \tilde{\boldsymbol{\omega}}_{ib}^b - \mathbf{C}_b^n \tilde{\boldsymbol{\omega}}_{in}^b\end{aligned}$$

Using Lemma 1.2 (Eq. (1.25)), one can see that

$$\mathbf{C}_b^n \tilde{\boldsymbol{\omega}}_{in}^b = \mathbf{C}_b^n \mathbf{C}_n^b \tilde{\boldsymbol{\omega}}_{in}^n \mathbf{C}_b^n = \tilde{\boldsymbol{\omega}}_{in}^n \mathbf{C}_b^n$$

Combining the above two equation gives

$$\dot{\mathbf{C}}_b^n = \mathbf{C}_b^n \tilde{\boldsymbol{\omega}}_{ib}^b - \tilde{\boldsymbol{\omega}}_{in}^n \mathbf{C}_b^n \tag{5.7}$$

With knowledge of the initial conditions of the transformation matrix, $\mathbf{C}_b^n$ can be solved as a function of time and hence the attitude of the craft relative to the Earth can be determined.

5.5 The Navigation Equation

Herein we will be using the vector dynamic equations developed in Appendix E. Suppose that **r** is the position relative to the inertial space of a navigating particle. As discussed before, its velocity in the inertial space is related to its velocity in an Earth fixed frame by [1]

$$\left.\frac{d\mathbf{r}}{dt}\right|_i = \left.\frac{d\mathbf{r}}{dt}\right|_e + \boldsymbol{\omega}_{ie} \times \mathbf{r} \tag{5.8}$$

In the above equation, the LHS term is the particle velocity as observed in the inertial space and the first term in the RHS is its velocity as observed in the Earth and shall be denoted by **v**, that is

$$\mathbf{v} = \left.\frac{d\mathbf{r}}{dt}\right|_e \tag{5.9}$$

To find accelerations we take the derivatives in the inertial space of every term in Eq. (5.8) (ignoring Earth angular acceleration) to get

$$\left.\frac{d^2\mathbf{r}}{dt^2}\right|_i = \left.\frac{d\mathbf{v}}{dt}\right|_i + \boldsymbol{\omega}_{ie} \times \left.\frac{d\mathbf{r}}{dt}\right|_i \tag{5.10}$$

Substituting for the inertial velocity from Eq. (5.8) in the above gives

$$\left.\frac{d^2\mathbf{r}}{dt^2}\right|_i = \left.\frac{d\mathbf{v}}{dt}\right|_i + \boldsymbol{\omega}_{ie} \times (\mathbf{v} + \boldsymbol{\omega}_{ie} \times \mathbf{r}) \tag{5.11}$$

Notice that the term on the LHS is the inertial acceleration of the vehicle in the inertial space. The first term in the RHS is the acceleration of the craft relative to the Earth as observed in the inertial space. This term must be converted to a frame relevant to our navigation purposes such as the Earth or the navigation frames. Implementing either of these two choices has its benefits. For example, the navigation frame computes the craft's position and velocity in terms of relevant quantities such as latitude, longitude and elevation. Alternatively, the Earth frame computes these variables in an Earth-fixed Cartesian frame. While this frame may be less convenient, it will be of value when interfacing with external navigation aids such as the global positioning system.

We start with the Nav-frame

$$\left.\frac{d\mathbf{v}}{dt}\right|_i = \frac{d\mathbf{v}}{dt} + \boldsymbol{\omega}_{in} \times \mathbf{v} \tag{5.12}$$

Notice that the first term in the RHS of the above equation is the craft's acceleration as observed in the Nav-frame. Substituting Eq. (5.12) in Eq. (5.11) and collecting terms gives

$$\left.\frac{d^2\mathbf{r}}{dt^2}\right|_i = \frac{d\mathbf{v}}{dt} + (\boldsymbol{\omega}_{in} + \boldsymbol{\omega}_{ie}) \times \mathbf{v} + \boldsymbol{\omega}_{ie} \times \boldsymbol{\omega}_{ie} \times \mathbf{r} \tag{5.13}$$

Substituting

$$\boldsymbol{\omega}_{in} = \boldsymbol{\omega}_{ie} + \boldsymbol{\omega}_{en} \tag{5.14}$$

in Eq. (5.13) and rearranging gives

$$\frac{d\mathbf{v}}{dt} = \left.\frac{d^2\mathbf{r}}{dt^2}\right|_i - (\boldsymbol{\omega}_{en} + 2\boldsymbol{\omega}_{ie}) \times \mathbf{v} - \boldsymbol{\omega}_{ie} \times \boldsymbol{\omega}_{ie} \times \mathbf{r} \tag{5.15}$$

Substituting the acceleration term of Eq. (5.5) in Eq. (5.15) we get

$$\frac{d\mathbf{v}}{dt} = \mathbf{a} - (\boldsymbol{\omega}_{en} + 2\boldsymbol{\omega}_{ie}) \times \mathbf{v} + \mathbf{g} \tag{5.16}$$

Again we emphasize that for vector sum consistency, the above variable must be represented in the same frame. For our purposes, the variables in Eq. (5.16) are now expressed in the Nav-frame since it is the one in which we desire to compute the position and velocity, and hence the equation becomes

$$\frac{d\mathbf{v}^n}{dt} = \mathbf{a}^n - (\boldsymbol{\omega}_{en}^n + 2\boldsymbol{\omega}_{ie}^n) \times \mathbf{v}^n + \mathbf{g}^n \tag{5.17}$$

This is the main navigation equation, a nonlinear differential equation in $\mathbf{v}$. Integrating it with respect to time will determine the craft velocity; integrating it once more will give the craft position. But before we do that we must determine the terms $\mathbf{a}^n$, $\mathbf{g}^n$, $\boldsymbol{\omega}_{en}^n$ and $\boldsymbol{\omega}_{ie}^n$. This is explained in the following remarks:

1. $\mathbf{a}^n$ is the accelerometer measurement vector projected in the Nav-frame. Given that accelerometer output vector $\mathbf{a}^b$ is measured in the body frame, $\mathbf{a}^n$ can be computed using the transformation matrix $\mathbf{C}_b^n$ given in Eq. (5.7)

$$\mathbf{a}^n = \mathbf{C}_b^n \mathbf{a}^b \tag{5.18}$$

2. $\mathbf{g}^n$ is the gravity vector term expressed in the Nav-frame. It is usually substituted for by the normal gravity (in section 4.8) as a function of the craft latitude and height relative to the ellipsoid.
3. $\boldsymbol{\omega}_{en}^n$ is the craft rate in the Nav-frame and is completely defined in section 4.3 and finally,
4. $\boldsymbol{\omega}_{ie}^n$ is the Earth rate in the Nav-frame, defined in section 4.4.

5.6 Navigation Equations Computational Flow Diagram

The following diagram depicts the computational flow of the navigation equations. We notice that the input data to this diagram are the gyro and accelerometer measurements in the body frame. The Earth spin rate vector $\boldsymbol{\omega}_{ie}^{e}$ is assumed a constant input. The gravity vector is shown as an input, which would be constant if it does not vary significantly, otherwise it will be computed as a function of height and Earth latitude. Finally, we remark that the data flow in this diagram merely shows the variable interdependence but not necessarily the actual computation in the navigation computer.

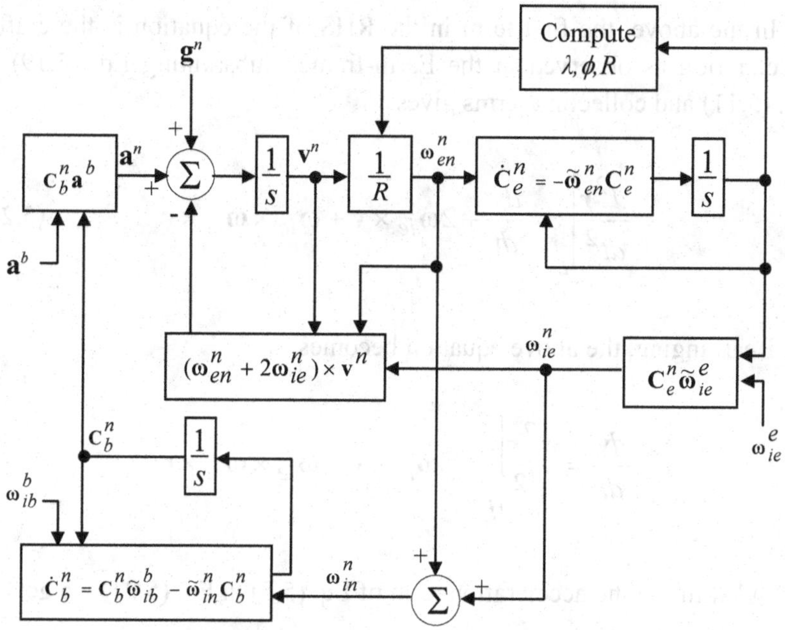

Figure 5.1 Navigation Equations Flow Diagram

5.7 The Navigation Equation in Earth Frame

The navigation equation in (5.17), when solved, provides the position and velocity in the navigation frame. Alternatively this equation could be derived so that these parameters are represented in Earth-Centered Earth-Fixed (ECEF) frame. As mentioned earlier this equation will be beneficial when interfacing with GPS. To derive the equations we start by writing Eq. (5.12) in Earth frame

$$\left.\frac{d\mathbf{v}}{dt}\right|_i = \frac{d\mathbf{v}}{dt} + \boldsymbol{\omega}_{ie} \times \mathbf{v} \tag{5.19}$$

In the above, the first term in the RHS of the equation is the craft's acceleration as observed in the Earth-frame. Substituting Eq. (5.19) in Eq. (5.11) and collecting terms gives

$$\left.\frac{d^2\mathbf{r}}{dt^2}\right|_i = \frac{d\mathbf{v}}{dt} + 2\boldsymbol{\omega}_{ie} \times \mathbf{v} + \boldsymbol{\omega}_{ie} \times \boldsymbol{\omega}_{ie} \times \mathbf{r} \tag{5.20}$$

Rearranging, the above equation becomes

$$\frac{d\mathbf{v}}{dt} = \left.\frac{d^2\mathbf{r}}{dt^2}\right|_i - 2\boldsymbol{\omega}_{ie} \times \mathbf{v} - \boldsymbol{\omega}_{ie} \times \boldsymbol{\omega}_{ie} \times \mathbf{r} \tag{5.21}$$

Substituting the acceleration term of Eq. (5.5) in Eq. (5.21) we get

$$\frac{d\mathbf{v}}{dt} = \mathbf{a} - 2\boldsymbol{\omega}_{ie} \times \mathbf{v} + \mathbf{g} \tag{5.22}$$

For vector sum consistency, the above variable must be represented in the same frame. Here, the variables in Eq. (5.22) are now expressed in the Earth-frame, and hence

$$\frac{d\mathbf{v}^e}{dt} = \mathbf{a}^e - 2\boldsymbol{\omega}_{ie}^e \times \mathbf{v}^e + \mathbf{g}^e \tag{5.23}$$

This chapter derived the navigation equation, Eq. (5.17), which computes the velocity and position in the navigation frame. Equation (5.23) performs the same computation in the Earth frame. The later equation will be needed when discussing inertial aided systems using the GPS. Using the computational flow chart in Fig. (5.1), we could proceed and solve the equation to obtain the desired parameters. We shall soon find that not all the variables in the equation change with time as rapidly as others. For example the sensor data are sampled at very high rates of 1 kHz or faster. In the mean time we see that a Boeing 747-400 cruising at a speed of 583 mph barely clears its footprints in a little over .25 seconds. This shows that not all the parameters need to be processed at the same rate. To take advantage of the higher rate data along the medium and slow rate parameters is the subject of the next chapter.

References

1. GR Pitman, Jr. (editor), Inertial Guidance, John Wiley and Sons Inc, New York, 1962

Chapter 6

Implementation

6.1 Introduction

The computational flow diagram in Chapter 5 reveals three sets of differential equations that need to be integrated and initialized. These sets are the craft attitude (inherent in $\mathbf{Q}_b^n$ or $\mathbf{C}_b^n$), the craft velocity, $\mathbf{v}$, and the craft position, inherent in $\mathbf{C}_e^n$. Of these variables, only $\mathbf{C}_e^n$ that does not depend explicitly on sensor data. Usually a conflict arises between high accuracy demands and limited computer resources when implementing integration algorithms. What we shall encounter is that inertial sensors supply data at high sampling rates, (typically at 1000 Hz or higher), while the variables that need to be integrated do not vary appreciably during these small time intervals. For example it suffices to compute the craft attitude at a rate of about 100 Hz and to compute the craft velocity and position at rate of about 10 Hz. To meet high accuracy requirements one might be tempted to integrate the navigation equations at the sensor high sampling rates. Doing so will inflicts a severe burden on computer resources. Conversely, integrating the equations at a slower rate, and not benefiting from the high sampling-rate data, will ultimately degrade the navigation solution.

Fortunately with little computational burden we can benefit from all the sensor data without overloading the navigation processor. We achieve that by preprocessing the sensor data at its high sampling rate and feeding the highly accurate solution to the navigation equations only at the attitude rate (100 Hz). The advantage of this preprocessing approach is that it can be implemented in an embedded sensor suite processor or, if not feasible, in the navigation processor without overloading it.

The preprocessing of sensor data will be implemented using the rotation vector method. Since this method will play a central part in our development, its solution will be provided in the following. We start with the computation of the two variables $\mathbf{C}_b^n$ and $\mathbf{v}$ that depend on the high sampling rate sensor data and discuss the rotation vector algorithm that will be instrumental in their integrations.

6.2 The Rotation Vector Differential Equation

Chapter 3 showed that the rotation vector [1] is one of the choices for computing frame transformation matrices and demonstrated its strong relevance to the quaternion. Even though the rotation vector comprises three elements in contrast of the four elements of the quaternion, its integration as seen in Eq. (3.51) is inconvenient and computationally prohibitive. However when the rotation angle is small the rotation vector differential equation simplifies considerably and becomes well suited for processing high rate sensor data. This equation as given in Eq. (3.52) is

$$\dot{\overline{\boldsymbol{\varphi}}} = \boldsymbol{\omega} + \frac{1}{2}\overline{\boldsymbol{\varphi}} \times \boldsymbol{\omega} \tag{6.1}$$

where $\overline{\boldsymbol{\varphi}}$ is the rotation vector and $\boldsymbol{\omega}$ is the input sensor data. It is apparent that solving for $\overline{\boldsymbol{\varphi}}$, when $\boldsymbol{\omega}$ (the deriving input to Eq. (6.1)) does not follow any temporal pattern and comes only in the form of sampled data could be a challenging problem. Therefore, solutions to the above equation were attempted by fitting $\boldsymbol{\omega}$ to 0th, 1st, 2nd or higher degree time polynomials over the integration time interval [2-6]. Naturally, the higher the degree of the polynomial the better the accuracy of the solution but it will be at the cost of a higher computational load. In the following, h is assumed to be the sampling time interval of $\boldsymbol{\omega}$.

Here we derive the solution assuming linearly varying sensor data [2]. In this case $\boldsymbol{\omega}$ can be represented by

$$\boldsymbol{\omega}(t) = \mathbf{a} + \mathbf{b}t \tag{6.2}$$

Using Taylor series expansion, the rotation vector, $\overline{\boldsymbol{\varphi}}$, can be expanded in time as follows

$$\overline{\boldsymbol{\varphi}}(h) = \overline{\boldsymbol{\varphi}}_0 + h\dot{\overline{\boldsymbol{\varphi}}}_0 + \frac{h^2}{2}\ddot{\overline{\boldsymbol{\varphi}}}_0 + \frac{h^3}{6}\dddot{\overline{\boldsymbol{\varphi}}}_0 + O(h^4) \tag{6.3}$$

where the subscripted parameters in Eq. (6.3) denote the rotation vector initial conditions at t=0. These initial conditions are obtained by deriving the higher derivatives of Eq. (6.1) and evaluating them at t=0. It will be assumed that the vector magnitudes of **a**, **b** and $\overline{\boldsymbol{\varphi}}_0$ are small that third order terms are ignored. Evaluating Eq. (6.2) at t=0 gives

$$\boldsymbol{\omega}_0 = \mathbf{a} \tag{6.4}$$

$$\dot{\boldsymbol{\omega}}_0 = \mathbf{b} \tag{6.5}$$

Using the above in Eq. (6.1) yields

$$\dot{\overline{\boldsymbol{\varphi}}}_0 = \mathbf{a} + \frac{1}{2}\overline{\boldsymbol{\varphi}}_0 \times \mathbf{a} \tag{6.6}$$

To determine $\ddot{\overline{\boldsymbol{\varphi}}}_0$ we take the derivative of Eq. (6.1) to get

$$\ddot{\overline{\boldsymbol{\varphi}}} = \dot{\boldsymbol{\omega}} + \frac{1}{2}\dot{\overline{\boldsymbol{\varphi}}} \times \boldsymbol{\omega} + \frac{1}{2}\overline{\boldsymbol{\varphi}} \times \dot{\boldsymbol{\omega}} \tag{6.7}$$

Evaluating Eq. (6.7) at t=0 and substituting from Eqs. (6.4)-(6.6) gives

$$\ddot{\overline{\varphi}}_0 = \mathbf{b} + \frac{1}{2}(\mathbf{a} + \frac{1}{2}\overline{\varphi}_0 \times \mathbf{a}) \times \mathbf{a} + \frac{1}{2}\overline{\varphi}_0 \times \mathbf{b}$$

Ignoring the higher order term in this equation yields

$$\ddot{\overline{\varphi}}_0 = \mathbf{b} + \frac{1}{2}\overline{\varphi}_0 \times \mathbf{b} \tag{6.8}$$

Also to determine $\dddot{\overline{\varphi}}_0$ we take the derivative of Eq. (6.7) to get

$$\dddot{\overline{\varphi}} = \ddot{\omega} + \frac{1}{2}\ddot{\overline{\varphi}} \times \omega + \dot{\overline{\varphi}} \times \dot{\omega} + \frac{1}{2}\overline{\varphi} \times \ddot{\omega} \tag{6.9}$$

Evaluating at t=0 and substituting from Eqs. (6.4)-(6.8) yields

$$\dddot{\overline{\varphi}}_0 = \frac{1}{2}(\mathbf{b} + \frac{1}{2}\overline{\varphi}_0 \times \mathbf{b}) \times \mathbf{a} + (\mathbf{a} + \frac{1}{2}\overline{\varphi}_0 \times \mathbf{a}) \times \mathbf{b}$$

Ignoring the third order terms in this equation gives

$$\dddot{\overline{\varphi}}_0 = \frac{1}{2}\mathbf{a} \times \mathbf{b} \tag{6.10}$$

Now substituting from Eqs. (6.6), (6.8) and (6.10) into Eq. (6.3) gives

$$\overline{\varphi}(h) = \overline{\varphi}_0 + h(\mathbf{a} + \frac{1}{2}\overline{\varphi}_0 \times \mathbf{a}) + \frac{h^2}{2}(\mathbf{b} + \frac{1}{2}\overline{\varphi}_0 \times \mathbf{b}) + \frac{h^3}{12}\mathbf{a} \times \mathbf{b} \tag{6.11}$$

Collecting terms, the above equation becomes

$$\overline{\varphi}(h) = \overline{\varphi}_0 + (h\mathbf{a} + \frac{h^2}{2}\mathbf{b}) + \frac{1}{2}\overline{\varphi}_0 \times (h\mathbf{a} + \frac{h^2}{2}\mathbf{b}) + \frac{h^3}{12}\mathbf{a} \times \mathbf{b} \quad (6.12)$$

It should be noted that for convenience, the using of t=0 (in the above equations to compute the initial condition) only meant to be a reference point to all time intervals. To determine **a** and **b** in Eq. (6.12) we have two options. The first [2] is to assume that $\boldsymbol{\omega}$ adhere to Eq. (6.2) over two sensor intervals, i.e. in the interval [0,2h]. At time 2h, Eq. (6.12) is

$$\overline{\varphi}(2h) = \overline{\varphi}_0 + 2(h\mathbf{a} + h^2\mathbf{b}) + \overline{\varphi}_0 \times (h\mathbf{a} + h^2\mathbf{b}) + \frac{2h^3}{3}\mathbf{a} \times \mathbf{b} \quad (6.13)$$

To determine the unknowns in the above equation let

$$\boldsymbol{\theta}_1 = \int_0^h \boldsymbol{\omega}(t)\, dt = \mathbf{a}t + \frac{1}{2}\mathbf{b}t^2 \Big|_0^h = h\mathbf{a} + \frac{1}{2}h^2\mathbf{b}$$

$$\boldsymbol{\theta} = \int_0^{2h} \boldsymbol{\omega}(t)\, dt = \mathbf{a}t + \frac{1}{2}\mathbf{b}t^2 \Big|_0^h = 2h\mathbf{a} + 2h^2\mathbf{b}$$

From which we get

$$\boldsymbol{\theta}_1 \times \boldsymbol{\theta} = (h\mathbf{a} + \frac{1}{2}h^2\mathbf{b}) \times (2h\mathbf{a} + 2h^2\mathbf{b}) = h^3\mathbf{a} \times \mathbf{b}$$

Consequently Eq. (6.13) becomes

$$\overline{\varphi}(2h) = \overline{\varphi}_0 + \boldsymbol{\theta} + \frac{1}{2}\overline{\varphi}_0 \times \boldsymbol{\theta} + \frac{2}{3}\boldsymbol{\theta}_1 \times \boldsymbol{\theta} \quad (6.14)$$

With the above option, we note that the rotation vector can be solved every other sensor data. However, in the second option [3], if we assume that $\boldsymbol{\omega}$ adhere to Eq. (6.2) over the interval [-*h*,*h*] then $(h\mathbf{a} + h^2/2\mathbf{b})$ and $h^3\mathbf{a}\times\mathbf{b}$ in Eq. (6.12) can be determined as follows:

$$\boldsymbol{\theta} = \int_0^h \boldsymbol{\omega}(t)\,dt = \mathbf{a}t + \frac{1}{2}\mathbf{b}t^2\Big|_0^h = h\mathbf{a} + \frac{1}{2}h^2\mathbf{b}$$

$$\boldsymbol{\theta}^p = \int_{-h}^0 \boldsymbol{\omega}(t)\,dt = \mathbf{a}t + \frac{1}{2}\mathbf{b}t^2\Big|_{-h}^0 = h\mathbf{a} - \frac{1}{2}h^2\mathbf{b}$$

$$\boldsymbol{\theta}^p \times \boldsymbol{\theta} = (h\mathbf{a} - \frac{1}{2}h^2\mathbf{b}) \times (h\mathbf{a} + \frac{1}{2}h^2\mathbf{b}) = h^3\mathbf{a}\times\mathbf{b} \tag{6.15}$$

where $\boldsymbol{\theta}^p$ and $\boldsymbol{\theta}$ are just two consecutive sampled outputs of the rate-integrating gyro. Substituting from the above equations into Eq. (6.12)

$$\overline{\boldsymbol{\varphi}}(h) = \overline{\boldsymbol{\varphi}}_0 + \boldsymbol{\theta} + \frac{1}{2}\overline{\boldsymbol{\varphi}}_0 \times \boldsymbol{\theta} + \frac{1}{12}\boldsymbol{\theta}^p \times \boldsymbol{\theta} \tag{6.16}$$

The subtle difference between Eq. (6.14) and (6.16) is due to the overlapping linearity. In Eq. (6.14) we assume that $\boldsymbol{\omega}$ follows Eq. (6.2) in the interval [0,2*h*] we compute the solution at time 2*h*, then we apply Eq. (6.2) in the interval [2*h*,4*h*] and compute the solution again at time 4*h*, and so on. Alternatively, in Eq. (6.16) we assume that $\boldsymbol{\omega}$ follows Eq. (6.2) in the interval [-*h*,*h*] and compute the solution at time *h*, then we apply Eq. (6.2) in the interval [0,2*h*] to compute the solution at time 2*h*, and so on.

Solving the rotation vector can be simplified if $\boldsymbol{\omega}(t)$ is assumed constant over the integration interval, *h*, equivalent to being fitted with a 0th degree polynomial. The solution to the rotation vector differential equation is obtained by setting **b** to **0** in Eq. (6.2) and Eq. (6.16) becomes

$$\bar{\varphi}=\bar{\varphi}_0+\theta+\frac{1}{2}\bar{\varphi}_0\times\theta \tag{6.17}$$

The constancy of $\omega(t)$ is a reasonable assumption, especially when:

1. The sensor sampling rate is high enough that better approximation may not yield significantly better accuracy.
2. The sensor error levels are of the same order as the higher order terms.
3. The desired accuracy does not warrant higher order approximation
4. The processor computational load is critical.

6.3 The Attitude Equation

The craft attitude is the orientation of the craft body relative to the navigation frame. The attitude equation can be updated using either the quaternion or the DCM. Starting with the quaternion and employing the notations used in Sec. 3.6.2, we obtain

$$\mathbf{Q}_n^b(t+T) = \mathbf{Q}_{n(t+T)}^{b(t+T)} = \mathbf{Q}_{n(t+T)}^{n(t)}\mathbf{Q}_{n(t)}^{b(t)}\mathbf{Q}_{b(t)}^{b(t+T)} = \mathbf{Q}_{n(t+T)}^{n(t)}\mathbf{Q}_n^b(t)\mathbf{Q}_{b(t)}^{b(t+T)} \tag{6.18}$$

The RHS of this equation contains two incremental quaternions, one for the body and one for the navigation, which must be computed to update the quaternion on the LHS. Both can be computed by integrating their respective quaternion differential equation or using the solution of the rotation vector. For its simplicity, the rotation vector has a clear is adopted for computing these terms.

For body frame computation, ω^b is the angular rate measured by the rate gyros. However rate gyros do not actually output the rate vector, but rather its integral over the short time interval in which it is sampled.

As such, they are very suitable for computing the solution of $\bar{\varphi}$ given by Eqs. (6.16) or (6.17). When one of these equations is selected to obtain the attitude solution it will be propagated for a number of time intervals, m, determined by the ratio of the attitude solution time interval, T, and the sensor sampling time interval, h, given as

$$m = \frac{T}{h} \tag{6.19}$$

To elaborate, let us consider the situation in which the sensor data are sampled at time intervals, h, of 1 ms and the attitude solution is needed at time intervals, T, of 10 ms Therefore at the beginning of an attitude cycle, the rotation vector must be computed and propagated 10 times, the same ratio of T to h. If, in a different situation, the solution of the attitude is desired at 1000 Hz (the same rate as the sensor) then $\boldsymbol{\varphi}$ is propagated only once. Hence the quaternion is also computed at 1000 Hz.

To proceed with the solution, let t be the instance at which the attitude solution is computed, then the rate gyro output at time $(t+ih)$ seconds, $(0<i \leq m)$, is given by

$$\boldsymbol{\theta}(t+ih) = \int_{t+ih-h}^{t+ih} \boldsymbol{\omega}^b(s)\,ds, \quad i = 1,2,\ldots,m \tag{6.20}$$

Supposing that the rate is constant during the sampling period, we can implement Eq. (6.17) to propagate the attitude solution and the equation becomes

$$\overline{\boldsymbol{\varphi}}(t+ih) = \overline{\boldsymbol{\varphi}}(t+ih-h) + \boldsymbol{\theta}(t+ih) + \frac{1}{2}\overline{\boldsymbol{\varphi}}(t+ih-h) \times \boldsymbol{\theta}(t+ih), \quad i = 1,2,\ldots,m \tag{6.21}$$

Alternatively if the rate varies linearly, we implement Eq. (6.16) to propagate the attitude solution, and the equation becomes

$$\begin{aligned}\bar{\boldsymbol{\varphi}}(t+ih) &= \bar{\boldsymbol{\varphi}}(t+ih-h)+\boldsymbol{\theta}(t+ih)\\ &+\left(\frac{1}{2}\bar{\boldsymbol{\varphi}}(t+ih-h)+\frac{1}{12}\boldsymbol{\theta}(t+ih-h)\right)\times\boldsymbol{\theta}(t+ih), \quad i=1,2,...,m\end{aligned} \tag{6.22}$$

Notice that attitude solution in Eqs. (6.21) and (6.22) are initialized with i=0 and $\bar{\boldsymbol{\varphi}}(t)=\mathbf{0}$.

At the end of m propagations, $\bar{\boldsymbol{\varphi}}(t+mh)$ from the above equation is used to compute the quaternion in Eq. (3.17), given by

$$\mathbf{Q}_{b(t)}^{b(t+T)} = \left\{\cos(\frac{\phi}{2}),\sin(\frac{\phi}{2})\frac{\bar{\boldsymbol{\varphi}}}{\phi}\right\} \tag{6.23}$$

To compute the incremental navigation quaternion, $\mathbf{Q}_{n(t+T)}^{n(t)}$, we have two options. The first is to compute the rotation vector at the same computational rate as that for the body. We follow exactly the same procedure we established for computing the body incremental rotation vector. Since the change of navigation frame with time is much slower than that of the body, the attitude vector solution in Eq. (6.17) is acceptable. When solving for the incremental navigation solution, $\boldsymbol{\omega}$ in Eq. (6.20) will refer to $\boldsymbol{\omega}_{in}^{n}$ which is determined by summing $\boldsymbol{\omega}_{ie}^{n}$ and $\boldsymbol{\omega}_{en}^{n}$ (given by Eqs. (4.7) and (4.11) respectively). Once the rotation vector is determined then $\mathbf{Q}_{n(t+T)}^{n(t)}$ is computed using Eq. (3.17). The other option is to compute the incremental navigation quaternion at a much slower rate, say every nT seconds. Hence Eq. (6.18) becomes

$$\begin{aligned}\mathbf{Q}_{n}^{b}(t+iT) &= \mathbf{Q}_{n}^{b}(t)\mathbf{Q}_{b(t)}^{b(t+iT)}, \quad i<n\\ &= \mathbf{Q}_{n(t+nT)}^{n(t)}\mathbf{Q}_{n}^{b}(t)\mathbf{Q}_{b(t)}^{b(t+nT)} \quad i=n\end{aligned}$$

Finally, it should be pointed out that the body and the navigation frame rotations in typical time intervals are very small. As such, a three term Taylor series expansion is very adequate to compute the sine and cosine terms in Eq. (6.23)

$$\cos\phi = 1 - \frac{\phi^2}{2} + \frac{\phi^4}{24}$$

$$\frac{\sin\phi}{\phi} = 1 - \frac{\phi^2}{6} + \frac{\phi^4}{120}$$

Had it been desired to use DCMs to update the attitude, one would then compute this equation

$$\mathbf{C}_b^n(t+T) = \mathbf{C}_{b(t+T)}^{n(t+T)} = \mathbf{C}_{n(t)}^{n(t+T)}\mathbf{C}_{b(t)}^{n(t)}\mathbf{C}_{b(t+T)}^{b(t)} = \mathbf{C}_{n(t)}^{n(t+T)}\mathbf{C}_b^n(t)\mathbf{C}_{b(t+T)}^{b(t)} \tag{6.24}$$

We would determine the rotation vectors for both the body and navigation frames exactly the same as shown in the quaternion solution. The incremental DCMs in Eq. (6.24) are then computed in terms of the rotation vector. For small rotational angles in which $\cos\phi \cong 1$ and $\sin\phi \cong \phi$, Eq. (3.7) becomes $\mathbf{C} = \mathbf{I} - \phi\tilde{\mathbf{u}}$ and from Eq. (3.6) we show that

$$\mathbf{C}_{b(t)}^{b(t+T)} = \mathbf{I} - \tilde{\boldsymbol{\varphi}}(t+T) \tag{6.25}$$

where $\tilde{\boldsymbol{\varphi}}$ is the skew symmetric matrix correspondent to the rotation vector in the time interval $[t, t+T]$. The incremental navigational DCM is computed similar to the body DCM and both are substituted into Eq. (6.24) to get the complete updated solution.

6.4 The Craft Velocity Equation

The craft velocity is the integral of its acceleration in the navigation frame. Craft acceleration is the vector sum of its inertial, Earth, Coriolos and Earth gravitational accelerations. The inertial acceleration is the most rapidly changing term and preprocessing it to the lower rates of the rest of the terms is demonstrated here.

Since the inertial acceleration in the navigation frame is

$$\mathbf{a}^n(t) = \mathbf{C}_b^n(t)\mathbf{a}^b(t) \tag{6.26}$$

We can integrate the above expression in the sensor time interval and substitute from Eq. (6.24) to give

$$\begin{aligned}
\Delta\mathbf{v}^n(t+ih) &= \int_{t+ih-h}^{t+ih} \mathbf{a}^n(s)\,ds \\
&= \int_{t+ih-h}^{t+ih} \mathbf{C}_b^n(s)\mathbf{a}^b(s)\,ds \\
&= \int_{t+ih-h}^{t+ih} \mathbf{C}_{n(t)}^{n(s)}\mathbf{C}_{b(t)}^{n(t)}\mathbf{C}_{b(s)}^{b(t)}\mathbf{a}^b(s)\,ds, \quad i=1,2,\ldots,m
\end{aligned} \tag{6.27}$$

Supposing that the sensor sample interval, h, is sufficiently small makes the incremental navigation DCM frame in the above practically a unit matrix and allows Eq. (6.27) to simplify to

$$\Delta\mathbf{v}^n(t+ih) = \mathbf{C}_{b(t)}^{n(t)} \int_{t+ih-h}^{t+ih} \mathbf{C}_{b(s)}^{b(t)}\mathbf{a}^b(s)\,ds, \quad i=1,2,\ldots,m \tag{6.28}$$

Further, if the body attitude is assumed constant during the sensor sample interval then the DCM inside the integral will be constant and Eq. (6.28) simplifies to

$$\Delta\mathbf{v}^n(t+ih) = \mathbf{C}_{b(t)}^{n(t)}\mathbf{C}_{b(t+ih)}^{b(t)} \int_{t+ih-h}^{t+ih} \mathbf{a}^b(s)\,ds, \quad i=1,2,\ldots,m \tag{6.29}$$

Let the sensor output be

$$\mathbf{v}(t+ih) = \int_{t+ih-h}^{t+ih} \mathbf{a}^b(s)\, ds \tag{6.30}$$

Substituting for the body DCM from Eqs. (6.25) and (6.30) into Eq. (6.29) results in

$$\begin{aligned}\Delta\mathbf{v}^n(t+ih) &= \mathbf{C}_{b(t)}^{n(t)}\mathbf{C}_{b(t+ih)}^{b(t)} \int_{t+ih-h}^{t+ih} \mathbf{a}^b(s)\, ds \\ &= \mathbf{C}_b^n(t)\left[\mathbf{I} + \tilde{\boldsymbol{\varphi}}(t+ih)\right]\mathbf{v}(t+ih) \\ &= \mathbf{C}_b^n(t)\left[\mathbf{v}(t+ih) + \overline{\boldsymbol{\varphi}}(t+ih)\times\mathbf{v}(t+ih)\right], \quad i = 1,2,\ldots,m\end{aligned} \tag{6.31}$$

In the above equation, we remark that:

1. The cross product is called the sculling term [6];
2. This equation is computed at the sensor data rate;
3. The rotation vector is obtained from Eq. (6.21).

If m is the number of sampled sensor data in each attitude computational interval, then the change in velocity in one attitude interval is given by

$$\mathbf{v}^n(t+T) = \mathbf{C}_b^n(t)\sum_{i=1}^{i=m}\left[\mathbf{v}(t+ih) + \overline{\boldsymbol{\varphi}}(t+ih)\times\mathbf{v}(t+ih)\right], \quad i = 1,2,\ldots,m \tag{6.32}$$

Had the rotation vector been varying linearly then the integral in Eq. (6.28) would be approximated by

$$\Delta\mathbf{v}^n(t+ih) = \mathbf{C}_{b(t)}^{n(t)}\left[\mathbf{I} + .5\tilde{\boldsymbol{\varphi}}(t+ih) + .5\tilde{\boldsymbol{\varphi}}(t+ih-h)\right]\mathbf{v}(t+ih), \quad i = 1,2,\ldots,m \tag{6.33}$$

The change in velocity in one attitude interval is the sum of the above equation over the m intervals, thus

$$\mathbf{v}^n(t+T) = \mathbf{C}_b^n(t)\sum_{i=1}^{i=m}\left[\mathbf{v}(t+ih) + .5\left(\overline{\boldsymbol{\varphi}}(t+ih) + \overline{\boldsymbol{\varphi}}(t+ih-h)\right)\times\mathbf{v}(t+ih)\right] \tag{6.34}$$

Alternatively, Eq. (6.34) can be computed by implementing the recurrence equation

$$\mathbf{v}^n(t+ih) = \mathbf{v}^n(t+ih-h) + \Delta\mathbf{v}^n(t+ih), \quad i = 1,2,\ldots,m \tag{6.35}$$

Like in the attitude computations, Eq. (6.35) is initialized with i=0 and $\overline{\boldsymbol{\varphi}}(t) = 0$ at the start of the attitude solution. The computational flow for implementing the attitude and velocity equations (6.22) and (6.35) is depicted in the diagram in Figure 6.1.

To summarize, the above procedure describes the preprocessing of the inertial sensor data. There are three different processing rates, namely the fast sensor data rate, the slow attitude rate and the much slower navigation rate. At the fast sensor sampling rate, sensor data are preprocessed and used to simultaneously compute and propagate the rotation vector using Eqs. (6.21) or (6.22), and the velocity increments in Eq. (6.33). At the slower attitude rate, the preprocessed data are used to compute the attitude quaternion or DCM and update the velocity to the navigation frame. Finally at the much slower navigation computational rate, the remaining Coriolos and gravity acceleration terms are integrated.

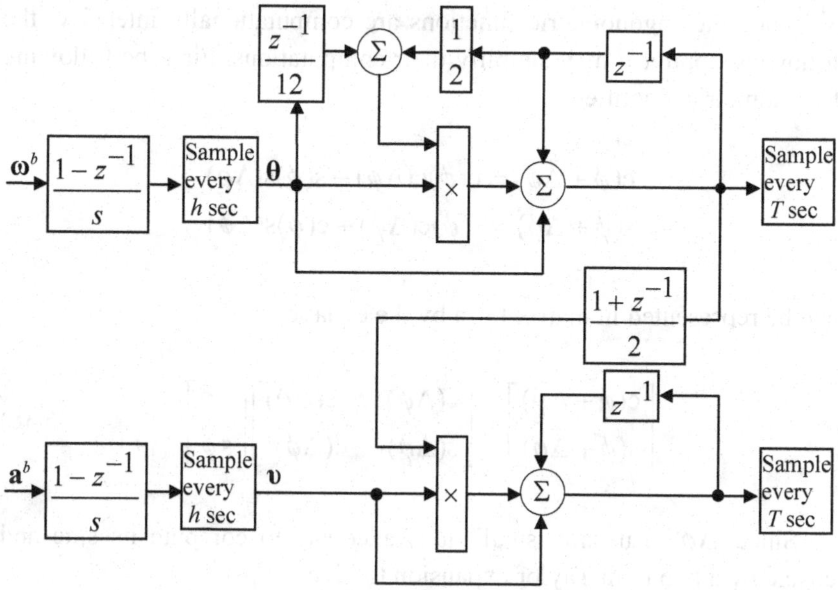

Figure 6.1 Inertial Sensors Computational Flow Diagram

6.5 The Craft Position Equation

The craft position determines its longitude, latitude and height relative to the ellipsoid. We could employ the procedure introduced in the previous section to integrate Eq. (4.14) to compute $\mathbf{C}_e^n$ and extract from it the longitude and latitude. This is unwarranted complication because we would be integrating nine elements to compute only two parameters, the latitude and longitude angles. Much easier is to integrate the longitude and latitude rates that are provided by Eqs. (4.8) and (4.9). With the updated longitude and latitude we can compute all the terms in $\mathbf{C}_e^n$.

We suppose that the navigation integration interval, T, is so small that the approximations in the following equations are accurate

$$\Delta\lambda = T\dot{\lambda} \tag{6.36}$$

$$\Delta\phi = T\dot{\phi} \tag{6.37}$$

Since the trigonometric functions are computationally intensive, the following approach may simplify their computations. First the following trigonometric identities

$$\begin{aligned} c(\phi+\Delta\phi) &= c(\phi)c(\Delta\phi) - s(\phi)s(\Delta\phi) \\ s(\phi+\Delta\phi) &= s(\phi)c(\Delta\phi) + c(\phi)s(\Delta\phi) \end{aligned}$$

can be represented in matrix form by the equation

$$\begin{bmatrix} c(\phi+\Delta\phi) \\ s(\phi+\Delta\phi) \end{bmatrix} = \begin{bmatrix} c(\Delta\phi) & -s(\Delta\phi) \\ s(\Delta\phi) & c(\Delta\phi) \end{bmatrix} \begin{bmatrix} c\phi \\ s\phi \end{bmatrix} \tag{6.38}$$

Since $\Delta\phi$ is usually small, it is adequate to compute its sine and cosine by a two term Taylor expansion to give

$$\begin{bmatrix} c(\phi+\Delta\phi) \\ s(\phi+\Delta\phi) \end{bmatrix} = \begin{bmatrix} 1-\dfrac{\Delta\phi^2}{2} & -\Delta\phi+\dfrac{\Delta\phi^3}{6} \\ \Delta\phi-\dfrac{\Delta\phi^3}{6} & 1-\dfrac{\Delta\phi^2}{2} \end{bmatrix} \begin{bmatrix} c\phi \\ s\phi \end{bmatrix} \tag{6.39}$$

A likewise longitude equation can be written as

$$\begin{bmatrix} c(\lambda+\Delta\lambda) \\ s(\lambda+\Delta\lambda) \end{bmatrix} = \begin{bmatrix} 1-\dfrac{\Delta\lambda^2}{2} & -\Delta\lambda+\dfrac{\Delta\lambda^3}{6} \\ \Delta\lambda-\dfrac{\Delta\lambda^3}{6} & 1-\dfrac{\Delta\lambda^2}{2} \end{bmatrix} \begin{bmatrix} c\lambda \\ s\lambda \end{bmatrix} \tag{6.40}$$

It is worth noting that the above two equations are numerically stable, as one can prove that the roots of the z-transform of each equation lie in the unit circle. The equations, when the angular increments are small enough, could be economically efficient provided that the cosine and sine terms are frequently normalized.

6.6 The Vertical Channel

So far we have shown how to implement the navigation equations that were derived in Chapter 5. In the past few sections, we discussed the means for integrating the variables that depend on the inertial sensor data. Herein we shall see that there is a unique problem associated with the vertical channel; direct integration will lead to numerical instability . This means that inertial sensors on their own can not be relied upon to compute valid vertical velocity or altitude. Here, we investigate the cause of this problem.

Recall, from Eq. (5.17), the navigation equation given by

$$\frac{d\mathbf{v}^n}{dt} = \mathbf{a}^n - (\boldsymbol{\omega}_{en}^n + 2\boldsymbol{\omega}_{ie}^n) \times \mathbf{v}^n + \mathbf{g}^n \tag{6.41}$$

To analyze the vertical channel we extract the z component in Eq. (6.41) to get

$$\begin{aligned}\frac{dv_z}{dt} &= \left[\mathbf{a}^n - (\boldsymbol{\omega}_{en}^n + 2\boldsymbol{\omega}_{ie}^n) \times \mathbf{v}^n + \mathbf{g}^n\right]_z \\ &= \left[\mathbf{a}^n - (\boldsymbol{\omega}_{en}^n + 2\boldsymbol{\omega}_{ie}^n) \times \mathbf{v}^n\right]_z + g_z\end{aligned} \tag{6.42}$$

Using the downward gravity given in Eq. (4.21) to substitute for g_z

$$g_z = g\left[1 - \frac{2}{a}\left(1 + f + m - 2f\sin^2(\phi)\right)h\right] \tag{6.43}$$

Ignoring the small terms in the above parentheses, and substituting in Eq. (6.42), it becomes

$$\frac{dv_z}{dt} = \left[\mathbf{a}^n - (\boldsymbol{\omega}_{en}^n + 2\boldsymbol{\omega}_{ie}^n) \times \mathbf{v}^n\right]_z + g\left(1 - \frac{2}{a}h\right) \tag{6.44}$$

To simplify the analysis let

$$A_z = -\left[\mathbf{a}^n - (\boldsymbol{\omega}_{en}^n + 2\boldsymbol{\omega}_{ie}^n) \times \mathbf{v}^n\right]_z - g \tag{6.45}$$

Recall that the navigation equations given in Eqs. (6.41) and (6.42) were developed for an NED frame where the z-axis points opposite to the positive direction of altitude h. This implies that

$$v_z = -\frac{dh}{dt} \tag{6.46}$$

Substituting for Eqs.(6.45) and (6.46) in (6.44) results in

$$\frac{d^2h}{dt^2} = A_z + \frac{2g}{a}h \tag{6.47}$$

Notice that:

1. The above equation is essentially what the computer does to propagate the altitude, h.
2. The equation is a second order differential equation in h.
3. The deriving input is A_z which, as observed from Eq. (6.45), is a combination of the sensed acceleration data, the Coriolos term and the gravity at nominally zero altitude.

To examine the numerical stability in the above equation we rearrange the terms to get

$$\frac{d^2h}{dt^2} - \frac{2g}{a}h = A_z \tag{6.48}$$

Applying the Laplace transform to the above differential equation

$$h = \frac{A_z}{s^2 - \frac{2g}{a}} \tag{6.49}$$

We can see that one of the roots of the characteristic equation in Eq. (6.49) is positive. This implies that the solution of the equation is numerically unstable. As such, this invalidates the use of the vertical channel of the navigation equation, Eq. (6.41). This problem is usually averted by employing an external aid for measuring the altitude like a radar altimeter or pressure altitude. The aiding data are integrated with the inertial data to get a stable and reliable solution to the vertical channel equation. More detail on computing the vertical channel, such as altitude and vertical speed, is discused in the next chapter.

References

1. Bortz, J.E., "A New Mathematical Formulation for Strapdown Inertial Navigation," IEEE Transactions on Aerospace and Electronic Systems, Vol. AES-7, No. 1, 1971, pp. 61-66.
2. Jordan, J.W., "An Accurate Strapdown Direction Cosine Algorithm," NASA TND-5384, Sept., 1969.
3. McKern, Richard A, "Study of Transformation Algorithm for Use in Digital Computer," MIT, Cambridge, Mass., Dept. of Aeronautics and Astronautics, M.S. Thesis, T493, Jan. 1968.
4. Miller, Robin B., "A New Strapdown Attitude Algorithm," J. Guidance, Vol. 6, No. 4, July-August 1983, pp. 287-291.
5. Y. F. Jiang and Y. P. Lin, "Improved Strapdown Coning Algorithms," IEEE Transactions on Aerospace and Electronic Systems, Vol. AES-28, No. 2, 1992.
6. V. Z. Gusinsky *et al*, "New Procedure for Deriving Optimized Strapdown Attitude Algorithms," Journal of Guidance, Control and Dynamics, Vol. 20, No. 4, 1997, pp. 673-680.
7. Bose, S., C. (2000): "Lecture Notes on GPS/INS Integrated Navigation Systems", Technalytics Inc., CA, USA.

Chapter 7

Air Data Computer

7.1 Introduction

Inertial navigation systems (INS) alone cannot be used for estimating altitude or vertical speed (rate of change of altitude) because of the associated numerical instability. Without an external aid, the computations of the INS vertical channel variables are bypassed and replaced with nominal values for altitude and vertical speed.

Alternatively, an air data system would complement the inertial navigation system for estimating the vertical channel variables. Since the early days of flight, air data systems have been part of aircraft navigation and virtually all aircraft are equipped with them. They have preceded the employment of inertial navigation systems as they provide crucial parameters – pressure altitude, vertical speed, density altitude and relative to wind airspeed – without which air navigation is impractical or unsafe. Thus we can see that the air data system complements and integrates well with the inertial navigation system.

Typically, an air data suite would comprise a free stream air pressure sensor, a pitot tube and an outside air temperature probe mounted on board of craft. The sensor data, when processed, can provide the craft altitude and the air speed. The free stream air pressure - also called the static pressure - is completely oblivious to the presence of any moving object and varies nonlinearly with the craft's altitude above ground. On the other hand, the air pressure that results from the air impact on the nose of a flying aircraft – called the stagnation pressure as it brings the air velocity to zero – is measured by the pitot tube. This pressure is also called the total pressure since it is the sum of the static pressure and the dynamic pressure that results from aircraft motion. Since the dynamic pressure is proportional to the square of the craft velocity then the difference between the total pressure and the static pressure can be used to compute the craft's speed.

In the following we will address the use of the air data system for navigation and how we interface it with the inertial navigation system to activate its vertical channel. It should be emphasized that at the core of an air data system are mere sensor data. Pressure data are converted to altitude using the concept of standard atmosphere. Based on some ideal assumptions the standard atmosphere provides a mathematical relationship between the outside air pressure and the altitude. First we introduce the standard, known as the "US Standard Atmosphere 1976", discuss the inherent assumptions and then derive the mathematical equations that relate the altitude to air pressure and vice versa. We will discuss how to estimate the craft's altitude and interface it with the vertical channel of the INS. Afterwards we will introduce and derive the equations for the density altitude, the vertical air speed, the air speed and the indicated airspeed as functions of the air pressure and temperature.

In this chapter we will come in touch with several types of altitudes: true altitude, geopotential altitude, pressure altitude and density altitude. For our purposes we assume that all these altitudes are referenced to the geoid, approximately the mean sea level (MSL). We just mention here that the true altitude (also called the geometric altitude) of a certain point is the actual (tape measure) altitude of this point above the MSL. We discuss the rest of altitudes in their respective locations.

7.2 US Standard Atmosphere 1976

This standard divides the Earth's atmosphere into layers of various heights, in each the air temperature varies according to a parameter called the lapse rate (*LR*) [1-2]. The lapse rate is the rate of change in temperature with respect to height throughout the layer. Table 7.1, adopted from [1], lists these layers and the temperature variations. For example, from the ground up to an altitude of 11 km, the air temperature decreases linearly with altitude at a rate of 6.5 deg K/ km. Notice that the pressure unit in the table is Pascal (Pa); a Pascal is a Newton/m^2. The height in the table is a hypothetical height known as geopotential height. This geopotential height amounts to the potential energy for lifting a unit mass to this altitude assuming a constant gravity field.

Table 7.1 Standard Atmosphere Layer Properties

Geopotential Height (m)	Temperature (K)	Lapse Rate (K/m)	Pressure (Pa)	Density (kg/m^3)
0	288.15	-0.0065	101325. 00	1.225
11000	216.65	0	22632.06	0.364
20000	216.65	+0.0010	5474.89	8.803E-02
32000	228.65	+0.0028	868.02	1.322E-02
47000	270.65	0	110.91	1.428E-03
51000	270.65	-0.0028	066.94	8.616E-04
71000	214.65	-0.0020	003.96	6.421E-05
84852	186.95		000.37	6.958E-06

In a layman terms, suppose we want to lift a unit mass (one Kg) to an altitude of say 1001 meters above MSL. How much potential energy would we inject into this mass? If gravity is constant everywhere, we would inject a potential energy of 1001 kg.m into the mass. In reality, the gravity field changes inversely proportional to the distance squared from the center of the earth. Therefore we would expect to inject into this mass 1000 kg.m (an exaggerated number) but nonetheless less potential energy. This expended potential energy is what we call the geopotential height. It has been shown that the relation between the geopotential altitude, *Z*, and the true altitude, *H*, is

$$Z = \frac{R\,H}{R + H}$$

where R is the radius of Earth [3]. We note that if the gravity is constant everywhere, the geopotential altitude will coincide with the true altitude.

The geopotential altitude concept is used to simplify the calibration of pressure sensors so they have the same reading at the same height anywhere on the globe. Assumptions incorporated in the standard atmosphere are:

1. at mean sea level air pressure is 101325 Pa.
2. at mean sea level air temperature is 288.15 K and
3. air is dry (humidity is zero).

These assumptions can hardly be met anytime or anywhere, and therefore corrections are made to compensate for deviation in assumptions.

7.3 Pressure Altitude

Here we derive the mathematical equations of air pressure vs. altitude to enable the conversion of sensed air pressure to altitude. In so doing, we shall use the geopotential height rather than the true height. This will imply that the gravity term, as will be shown, to be constant at the value at the mean sea level and hence will facilitate the deriving of the equations.

As depicted in Fig. 7.1, we consider an infinitesimal cylindrical shaped air mass of uniform cross sectional area, A, and height, dZ, at a certain (geopotential) height, Z, above sea level

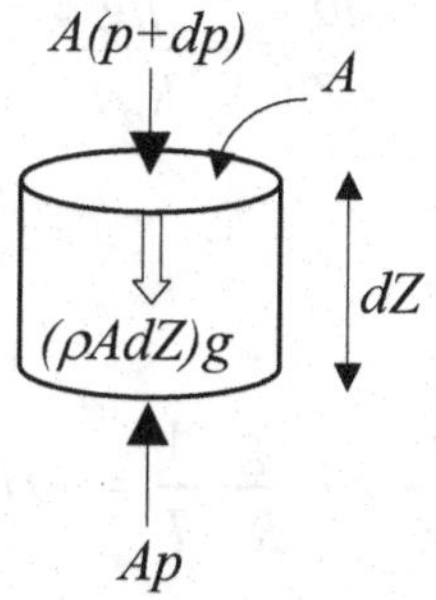

Figure 7.1 Equilibrium of forces on an infinitesimal air cylinder

Let us consider its vertical force equilibrium [3-4]. The gravity attraction on the cylinder mass balances with the air pressures acting on the cylinder bases, hence

$$(\rho AdZ)g + A(p+dp) = Ap$$

where g (9.80665 m/s^2) is the gravity constant at mean sea level, ρ is the air density and dp is air pressure difference on the cylinder. The above is simplified to the hydrostatic equation given by

$$\frac{dp}{dZ} = -\rho g \tag{7.1}$$

Assuming that air is dry and follows the perfect gas equation, $p = \rho R_d T$, implies that

$$\rho = \frac{p}{R_d T} \tag{7.2}$$

where R_d (287.05307 J /kg K) is the gas constant for dry air. Substituting for ρ from Eq. (7.2) into Eq. (7.1) yields

$$\frac{dp}{dZ} = -\frac{pg}{R_d T}$$

which implies that

$$\frac{1}{p}\frac{dp}{dZ} = -\frac{g}{R_d}\frac{1}{T} = -H_c \frac{1}{T} \tag{7.3}$$

where

$$H_c = \frac{g}{R_d} = .0341631947 \text{ K/m} \tag{7.4}$$

is the hydrostatic constant. The temperature, T, must be known in order to integrate and solve Eq. (7.3). From Table 7.1 we observe that T follows two patterns: the first is in which the temperature varies linearly with height (lapse rate is nonzero) and the second is in which the temperature is held constant (lapse rate is zero). When the lapse rate is not zero, the temperature, T, varies linearly with height according to

$$T(Z) = T_i + LR_i\left(Z - Z_i\right), \quad LR_i \neq 0 \tag{7.5}$$

where Z_i, T_i and LR_i are the height, temperature and lapse rate at the ith atmosphere layer level at which the lapse rate is nonzero. Substituting for T from Eq. (7.5) into Eq. (7.3) and integrating gives

$$p(Z) = P_i\left[1 + \frac{LR_i}{T_i}\left(Z - Z_i\right)\right]^{\left(-\frac{H_c}{LR_i}\right)}, \quad LR_i \neq 0 \tag{7.6}$$

Alternatively in the layer where the lapse rate is zero, $T(Z)$ is constant and Eq. (7.3) integrates to

$$p(Z) = P_i \exp\left[\frac{-H_c\left(Z - Z_i\right)}{T_i}\right], \quad LR_i = 0 \tag{7.7}$$

From Eq. (7.6), and using the constants from the first layer in Table 7.1, we get

$$p(Z) = 101325\left(1 - \frac{Z}{44330.77}\right)^{-5.25588} \text{ Pa}, \quad 0 \leq Z \leq 11000 \text{ m} \tag{7.8}$$

Similarly, from Eq. (7.7) and using the constants from the second layer in Table 7.1 we get

$$p(Z) = 22632.06 \exp\left(\frac{11000 - Z}{63416.62}\right) \text{Pa}, \quad 11000\,\text{m} \le Z \le 20000\,\text{m} \quad (7.9)$$

Conversely, the pressure altitude as a function of pressure is given by the following equations. For non zero lapse rates the altitude is given by

$$Z(p) = Z_i + \left(\frac{T_i}{LR_i}\right)\left[\left(\frac{p}{P_i}\right)^{-\frac{LR_i}{H_c}} - 1\right], \quad LR_i \neq 0 \qquad (7.10)$$

and for zero lapse rate the altitude is given by

$$Z(p) = Z_i - \frac{T_i}{H_c}\ln\left[\frac{p}{P_i}\right], \quad LR_i = 0 \qquad (7.11)$$

From Eq. (7.10), and using the constants from the first layer in Table 7.1, we get

$$Z(p) = 44330.77\left[1 - \left(\frac{p}{101325}\right)^{0.190263}\right] \text{m}, \qquad (7.12)$$

$$22632.06 \text{ Pa} \le p \le 101325 \text{ Pa}$$

Similarly, from Eq. (7.11), and using the constants from the second layer in Table 7.1, we get

$$Z(p) = 11000 + 6341.62 \ln\left[\frac{22632.06}{p}\right] \text{m}, \tag{7.13}$$

$$5474.89 \text{ Pa} \le p \le 22632.06 \text{ Pa}$$

In terms of the true altitude the above equations could be modified in two ways. The first in which (after computing the geographical altitude *Z*) the true altitude is determined from the relation $H = RZ/(R-Z)$. The second is by deriving the above equations in terms of the true altitude rather than the geopotential height: modify Eq. (7.1) by replacing the geopotential height, *Z*, with the true altitude, *H*, and the gravity constant, *g*, with the variable gravity term $gR^2/(R+H)^2$. This will lead to exactly equivalent equations as those given above.

7.4 Vertical Channel Parameter Estimation Using Inertial and Air Data

The previous chapter demostrated that the inertial vertical channel is numerically unstable and consequently would yield unreliable altitude estimates. With minor differences, several mechanizations have been reported (Fig. 10.2 [5], Fig. 7.14 [6], and Fig. 1 [7]) to correct this problem. The one proposed in [5] is probably oriented towords space applications while those in [6-7] are for aircraft applications. The approach, in these methods, is to combine the pressure altitude with the inertial data using a digital filter that provide a stable estimate of the altitude. By virtue of this methodology, the fast response inertial data (that alone would not be able to provide stable estimates) will be utilized to minimize the effects of the typically slower response pressure altitude data.

In the following, we present the main theme of the altitude estimation algorithm given in [7]. This algorithm, which blends the inertial data and the barometric measurements through a linear filter, is depicted in the block diagram in Fig. 7.2.

In the block diagram, *H* is the altitude reference derived from air data, *h* is the altitude estimate and *s* is the Laplace operator.

The inertial data term, given in Chapter 6 by Eq. (6.45), is

$$A_z = -\left[\mathbf{a}^n - (\boldsymbol{\omega}_{en}^n + 2\boldsymbol{\omega}_{ie}^n) \times \mathbf{v}^n\right]_z - g \tag{7.14}$$

From the diagram one can obtain

$$h = \frac{1}{s^2}\left(A_z + \frac{2g}{a}h\right) + \left(\frac{c_1}{s} + \frac{c_2}{s^2} + \frac{c_3}{s^3}\right)(H - h)$$

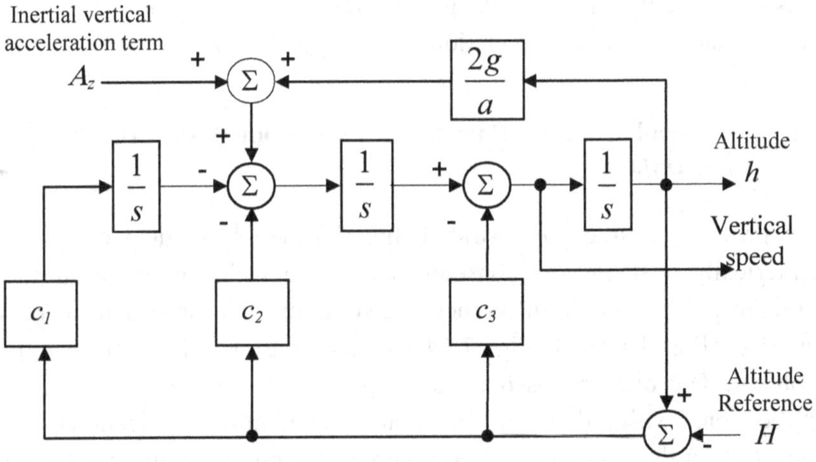

Figure 7.2 Vertical Channel Block Diagram

Collecting the h terms on one side gives

$$\left(1 + \frac{c_1}{s} + \frac{c_2 - \frac{2g}{a}}{s^2} + \frac{c_3}{s^3}\right)h = \frac{1}{s^2}A_z + \left(\frac{c_1}{s} + \frac{c_2}{s^2} + \frac{c_3}{s^3}\right)H \tag{7.15}$$

and implies that the characteristic equation of the computational filter is

$$1+\frac{c_1}{s}+\frac{c_2-\frac{2g}{a}}{s^2}+\frac{c_3}{s^3}=0 \tag{7.16}$$

Selecting the computational filter parameters to be

$$c_1=\frac{3}{\tau}$$

$$c_2=\frac{3}{\tau^2}+\frac{2g}{a}$$

$$c_3=\frac{1}{\tau^3}$$

will cause the roots of the characteristic equation to be equal and negative and hence ensures the numerical stability of the vertical channel. Typically, τ is selected to be 200 seconds [7].

In the above design, the filter is described in the continuous time domain and thus needs to be converted to a digital filter to allow for its implementation on a digital computer. The bilinear z- transform is amongst the most popular algorithms for performing this conversion [8-10]. In this method the Laplace operator, *s*, is replaced by

$$\frac{1}{s}=\frac{T}{2}\left(\frac{1+z^{-1}}{1-z^{-1}}\right)$$

where T is the filter computional time interval, and z is the delay operator.

What remains to implement the above algorithm is to determine H. Due to the incongruence of the standard atmospheric pressure altitude and the true altitude, attempting to substitute H with the pressure altitude in Eqs. (7.8) and (7.9) can lead to large errors in the altitude estimate [7]. The error in pressure altitude is quite substantial and could be as much as

5-10% of the true altitude and as such could degrade the INS vertical channel computations that are based on true altitude. There are at least two sources that cause the discrepancy between the pressure altitude and the true altitude. The first is that at sea level, the standard atmosphere is assumed to have a pressure of 101325 Pa and temperature of 288.15 K, an assumption that could rarely happen in real life. The second is that gravity is assumed constant at all altitudes and latitudes.

To counter these concerns, the algorithm adapts the hydrostatic equation to compute the true altitude rather than the pressure altitude derived from standard atmosphere. Substituting from Eq. (7.2) into Eq. (7.1) gives

$$dH = -\frac{R_d T}{g(\phi, H)p} dp$$

where, H is the true altitude, T and p are the measured temperature and pressure respectively, and g is the gravity. The gravity term, g, varies with latitude, ϕ, and height, H, and will be computed from

$$g(\phi, H) = g_0 + g_1 \sin^2(\phi) + g_2 H$$

where g_0, g_1 and g_2 are the first order approximation terms for computing the downward gravity component, γ_h in Eq. (4.21). The altitude is now propagated recursively as follows: Let the suffixes, *n-1* and *n*, denote the altitude solution in two consecutive intervals, and let D be defined as

$$D = \frac{T}{p}$$

Using the trapezoidal rule, the true altitude propagation equations become

$$D_n = \frac{T_n}{p_n}$$
$$g = g_0 + g_1 \sin^2(\phi) + g_2 H_{n-1}$$
$$H_n = H_{n-1} - \frac{R_d}{2g}(D_n + D_{n-1})(p_n - p_{n-1}) \tag{7.17}$$
$$p_{n-1} = p_n$$
$$D_{n-1} = D_n$$

Notice that both the pressure and the temperature are used in the above relations. The above algorithm is initialized with

$$H_0 = Alt\_init$$
$$D_0 = \frac{T_{Alt\_init}}{p_{Alt\_init}} \tag{7.18}$$
$$p_0 = p_{Alt\_init}$$

where *Alt_init*, in the above equation, is the initial altitude at the start of computations, and p_{Alt_init} and T_{Alt_init} are the pressure and temperature at that altitude.

Air sensor data are not only used to estimate the vertical channel parameters, but also to estimate other air parameters that are relevant to the performance of the aircraft, like the density altitude, the ascend/descend rate, the Mach number, air speed and indicated air speed. We address these parameters and their use in the following.

7.5 Density Altitude

Density altitude denotes the altitude at which air density attains a certain value. This parameter is not used for computing the craft altitude but rather is a measure of air density [11]. It turns out that it is easier measure the performance of the aircraft parameters in terms of density altitude than in actual density units. Some of these density dependent parameters are aerodynamic lift, drag, propeller thrust and engine horsepower. In the following we shall derive the density altitude as a function of the air sensor data. Equations (7.10)-(7.11) provide the altitude as a function of pressure. We derive similar equations that provide altitude as a function of density. From the universal gas equation, we have

$$\frac{\rho}{\rho_i} = \frac{p}{P_i}\frac{T_i}{T} \tag{7.19}$$

Consider the case of non zero lapse rate and notice from Eq. (7.6) that

$$\frac{p}{P_i} = \left(\frac{T}{T_i}\right)^{-\frac{H_c}{LR_i}} \tag{7.20}$$

Substituting from Eq. (7.19) into (7.20), gives

$$\frac{\rho}{\rho_i} = \left(\frac{T}{T_i}\right)^{-\frac{H_c}{LR_i}-1} \tag{7.21}$$

from which we get the inverse relation

$$\left(\frac{T}{T_i}\right)=\left(\frac{\rho}{\rho_i}\right)^{\frac{-1}{\frac{H_c}{LR_i}+1}} \qquad (7.22)$$

Substituting from (7.5), we get the density altitude as

$$Z(\rho)=Z_i+\frac{T_i}{LR_i}\left[\left(\frac{\rho}{\rho_i}\right)^{\frac{-1}{\frac{H_c}{LR_i}+1}}-1\right] \qquad (7.23)$$

Thus to compute the density altitude we compute the air density from the static pressure and temperature, and then substitute in Eq. (7.23).

7.6 Altitude (Descend /Climb) Rate

In absence of inertial navigation data, altitude rate can be obtained by differentiating the pressure altitude using a, commonly named, washout filter. The analog transfer function of this filter is

$$H(s)=\frac{s}{(1+\tau_1 s)(1+\tau_2 s)} \qquad (7.24)$$

The time constants in the above transfer function are selected so that $S_r \le (1/\tau_1 \text{ and } 1/\tau_2) \le N$, where S_r is the data computational rate, and N is the sensor noise bandwidth.

7.7 Air Speed

The equation that relates air pressures to the Mach number (the ratio of air speed to speed of sound) is given by

$$M^2 = \frac{2}{\gamma - 1}\left[\left(\frac{p_t}{p_s}\right)^{\frac{\gamma-1}{\gamma}} - 1\right] \tag{7.25}$$

where γ is the specific gas ratio constant. The speed of sound equation is given by

$$a^2 = \gamma R_d T_s \tag{7.26}$$

where T_s is the static air temperature and is related to the measured total air temperature, T_t, by

$$T_s = \frac{T_t}{1 + \frac{\gamma - 1}{2} M^2} \tag{7.27}$$

By computing the Mach number from Eq. (7.25), the static air temperature from Eq. (7.27) and the air speed from Eq. (7.26), we can compute the air speed by the equation

$$v = aM \tag{7.28}$$

This equation determines the craft speed relative to wind. Since inertial speed is relative to ground, the difference between these two speeds determines the wind velocity.

7.8 Indicated Air Speed (IAS)

The indicated air speed retains the same value for certain aerodynamic performance parameters and hence its importance [12]. For example, if a craft has an indicated stalling speed at 62 knots, it will remain the same at sea level or at 10,000 ft altitude. It is computed by the equation

$$IAS = \sqrt{\frac{2}{\rho_{SL}}(p_t - p_s)} \tag{7.29}$$

where ρ_{SL} is the air density at sea level. Derivation of the speed equations is given in Appendix F.

Even though we can solve the navigation equation almost everywhere there are areas in which that this cannot be done: the Polar Regions. There are mathematical problems associated with these areas that arise mainly because of the implementation of the spherical coordinates which use the latitude and longitude grid to locate objects. Specifically, in these areas all the longitude lines converge on at one point. Addressing this issue is the subject of the next chapter.

References

1. MJ Mahoney, The US Standard Atmosphere 1976, http://mtp.jpl.nasa.gov/notes/altitude/StdAtmos1976.html
2. John D. Anderson, Fundamentals of Aerodynamics, McGraw Hills, NY, New York, 1991
3. MJ Mahoney, A Discussion of Various Measures of Altitude http://mtp.jpl.nasa.gov/notes/altitude/altitude.html
4. L.M. Milne-Thompson, Theoretical Aerodynamics, Macmillan &Co LTD, 1958
5. M. Fernandez and G. Macomber, Inertial Navigation Engineering, Prentice Hall, Englewood Cliffs, New Jersey, 1962.
6. M. Kayton and W. Fried, editors, Avionics Navigation Systems, John wiley & sons, NY, New York, 1969.
7. R.L. Blanchard, "A new algorithm for computing inertial attitude and velocity," IEEE Trans. Aerospace Electronic Systems, 1971, AES-7, 1143-1176.
8. Ifeachor, I.C. and Jervis, B.W., Digital Signal Processing, Addison Wesley, 1993.

9. A. V. Oppenheim, R. W. Schafer, Digital Signal Processing, Prentice Hall, Inglewood Cliffs, New Jersey, 1975.
10. Papoulis, Signal Analysis, McGraw Hill, New York, 1977.
11. Richard Shelquist, "Air Density and Density Altitude Calculations," Equations - Air Density and Density Altitude
12. Hal Stoen, Understanding Airspeed, 6/19/2000, http://stoenworks.com/Tutorials/Understanding%20airspeed.html

Chapter 8

Polar Navigation

8.1 Introduction

Today polar navigation is a reality and no longer an adventure or an exploratory pursuit. Polar routes have been established and more airlines are navigating these routes [1]. Here we address some of the mathematical challenges associated with polar navigation and provide simple solutions.

The DCM from the Earth to the Nav-frame, in Eq. (4.14), is given by

$$\dot{\mathbf{C}}_e^n = -\tilde{\boldsymbol{\omega}}_{en}^n \mathbf{C}_e^n \tag{8.1}$$

where its parameters, given by Eqs. (4.6) and (4.11), are

$$\boldsymbol{\omega}_{en}^n = \begin{bmatrix} \rho_n \\ \rho_e \\ \rho_d \end{bmatrix} = \begin{bmatrix} \dfrac{v_e}{R_p + h} \\ -\dfrac{v_n}{R_m + h} \\ -\dfrac{v_e \tan\phi}{R_p + h} \end{bmatrix} = \begin{bmatrix} \dot{\lambda}\cos\phi \\ -\dot{\phi} \\ -\dot{\lambda}\sin\phi \end{bmatrix} \tag{8.2a}$$

$$\mathbf{C}_e^n = \mathbf{C}_y(-\phi)\,\mathbf{C}_x(\lambda) = \begin{bmatrix} \cos\phi & -\sin\phi\sin\lambda & \sin\phi\cos\lambda \\ 0 & \cos\lambda & \sin\lambda \\ -\sin\phi & -\cos\phi\sin\lambda & \cos\phi\cos\lambda \end{bmatrix} \tag{8.2b}$$

With this formulation, a potential problem will arise if the craft navigates close to the Earth's poles, when the latitude angle, ϕ, approaches ±90° and hence $\tan\phi$ will be infinite.

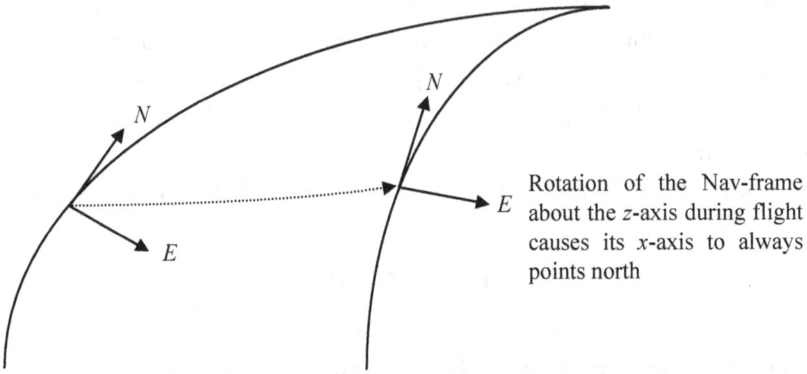
N
N
E
E
Rotation of the Nav-frame about the z-axis during flight causes its x-axis to always points north

Figure 8.1 North-East Nav-frame Orientation During Flight

Aside from the mathematical issues, let us see how this affects an actual flight path. We know that the longitudinal lines are not parallel lines and that all meet at the north and south poles. At any point in the navigation path, the vertical rotation rate of the craft, ρ_d, in Eq. (8.2), prescribes the amount of rotation that aligns the x and y axes of the Nav-frame to the north and east directions, respectively (see Fig. 8.1). Near the poles, the rate of change of the longitudinal angle increases rapidly, and consequently the Nav-frame must also rotate rapidly about the z-axis to keep itself aligned to the north direction.

8.2 The Wander Azimuth Navigation

The wander azimuth was the approach to circumvent the polar navigation problem. The idea is to force $\rho_d = 0$ to prevent the Nav-frame from rotating about the *z*-axis. Doing so sacrifices the elegance of computing in a Nav-frame aligned with the north and east axes. This will force the *x*-axis to align with the direction at which the flight is initialized. At any point in time the *x*-axis of the Nav-frame will deviate from the true north by an angle called the wander azimuth angle, α (see Fig. 8.2).

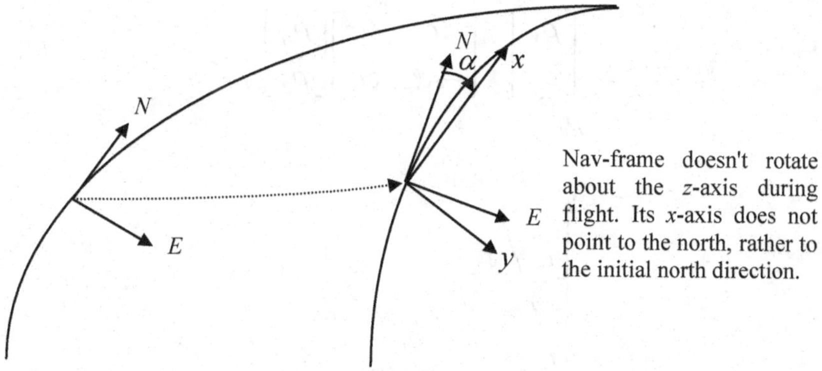
N
α
x
N
E
E
y
Nav-frame doesn't rotate about the z-axis during flight. Its x-axis does not point to the north, rather to the initial north direction.

Figure 8.2 Wander-Azimuth Nav-frame Orientation During Flight

In wander azimuth navigation the DCM of the Nav-frame will be modified by the wander azimuth angle, α, to be

$$\mathbf{C}_e^n = \mathbf{C}_z(\alpha)\,\mathbf{C}_y(-\phi)\,\mathbf{C}_x(\lambda)$$
$$= \begin{bmatrix} \mathrm{c}\alpha\,\mathrm{c}\phi & -\mathrm{c}\alpha\,\mathrm{s}\phi\,\mathrm{s}\lambda + \mathrm{s}\alpha\,\mathrm{c}\lambda & \mathrm{c}\alpha\,\mathrm{s}\phi\,\mathrm{c}\lambda + \mathrm{s}\alpha\,\mathrm{s}\lambda \\ -\mathrm{s}\alpha\,\mathrm{c}\phi & \mathrm{s}\alpha\,\mathrm{s}\phi\,\mathrm{s}\lambda + \mathrm{c}\alpha\,\mathrm{c}\lambda & -\mathrm{s}\alpha\,\mathrm{s}\phi\,\mathrm{c}\lambda + \mathrm{c}\alpha\,s\lambda \\ -\mathrm{s}\phi & -\mathrm{c}\phi\,\mathrm{s}\lambda & \mathrm{c}\phi\,\mathrm{c}\lambda \end{bmatrix} \tag{8.3}$$

In the above we have used the shorthand sα for sin(α) and likewise for the cosine. The associated craft rate will be

$$\boldsymbol{\omega}_{en}^{n} = \begin{bmatrix} \rho_x \\ \rho_y \\ 0 \end{bmatrix} \tag{8.4}$$

We now determine the craft rates in Eq. (8.4) in the new wander azimuth frame. As mentioned earlier, the new *x-y* plane will be deviating from the north-east by the wander azimuth angle as shown in Fig. 8.3. Therefore we can write

$$\begin{bmatrix} \rho_x \\ \rho_y \end{bmatrix} = \begin{bmatrix} c\alpha & s\alpha \\ -s\alpha & c\alpha \end{bmatrix} \begin{bmatrix} \rho_n \\ \rho_e \end{bmatrix} \tag{8.5}$$

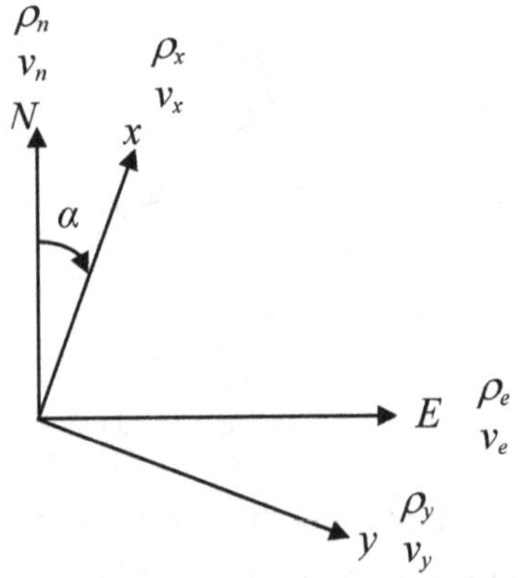

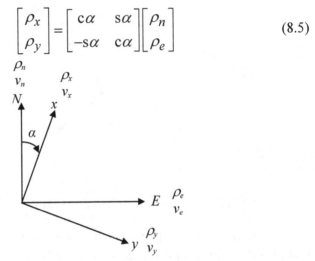

Figure 8.3 Velocities and Radii of Curvature in the Wander Azimuth Frame

In view of Fig. (8.3), Eq. (8.2) can be formulated in the new axes frame as

$$\begin{bmatrix} \rho_n \\ \rho_e \end{bmatrix} = \begin{bmatrix} 0 & \dfrac{1}{R_p + h} \\ \dfrac{-1}{R_m + h} & 0 \end{bmatrix} \begin{bmatrix} v_n \\ v_e \end{bmatrix}$$

which implies

$$\begin{bmatrix} \rho_n \\ \rho_e \end{bmatrix} = \begin{bmatrix} 0 & \dfrac{1}{R_p + h} \\ \dfrac{-1}{R_m + h} & 0 \end{bmatrix} \begin{bmatrix} c\alpha & -s\alpha \\ s\alpha & c\alpha \end{bmatrix} \begin{bmatrix} v_x \\ v_y \end{bmatrix} \tag{8.6}$$

Substituting from Eq. (8.6) into (8.5) yields

$$\begin{bmatrix} \rho_x \\ \rho_y \end{bmatrix} = \begin{bmatrix} c\alpha & s\alpha \\ -s\alpha & c\alpha \end{bmatrix} \begin{bmatrix} 0 & \dfrac{1}{R_p + h} \\ \dfrac{-1}{R_m + h} & 0 \end{bmatrix} \begin{bmatrix} c\alpha & -s\alpha \\ s\alpha & c\alpha \end{bmatrix} \begin{bmatrix} v_x \\ v_y \end{bmatrix}$$

Carrying out the matrix products in the above yields

$$\begin{bmatrix} \rho_x \\ \rho_y \end{bmatrix} = \begin{bmatrix} (\dfrac{1}{R_p + h} + \dfrac{1}{R_m + h}) s\alpha\, c\alpha & (\dfrac{c^2\alpha}{R_p + h} - \dfrac{s^2\alpha}{R_m + h}) \\ (\dfrac{-s^2\alpha}{R_p + h} + \dfrac{c^2\alpha}{R_m + h}) & -(\dfrac{1}{R_p + h} + \dfrac{1}{R_m + h}) s\alpha\, c\alpha \end{bmatrix} \begin{bmatrix} v_x \\ v_y \end{bmatrix}$$

(8.7)

Therefore to propagate the position DCM (given by Eq. (8.1)) in the wander azimuth frame, we must compute Eqs. (8.3) and (8.7) in place of Eq. (8.2). The above basically describes the needed modifications to implement the wander azimuth navigation algorithm.

8.3 Prospective of the Wander Azimuth Approach

One can see from Eqs. (8.3) and (8.7), as compared with Eq. (8.2), that the computational burden of the new approach is overwhelming. In the mean time, we lose direct track of the north-east navigation frame. And at this point, one might ponder the real need for the wander azimuth computations. Are they justified by any means? On the pro side, infinite longitudinal rate cannot be tolerated. On the con side, it is arguable that implementing this approach will result in much computational burden, most of which is dedicated to $\mathbf{C}_e^n$. To gain a perspective, consider this scenario in which a craft is circumnavigating a polar circle of radius r whose latitude is ϕ (the radius is then given by $r = R_p c\phi$). For simplicity, the height, h, is ignored. Then from Eq. (8.2)

$$\begin{aligned} \dot{\lambda}\, c\phi &= \frac{v_e}{R_p} \Rightarrow \\ r = R_p\, c\phi &= \frac{v_e}{\dot{\lambda}} \end{aligned} \tag{8.8}$$

Suppose the craft is traveling at the speed of sound (340 m/s^2) and that a change of 30° in longitudinal angle in a computational frame of .04 second is tolerated, then

$$r = \frac{v_e}{\dot{\lambda}} = \frac{v_e \Delta t}{\Delta \lambda} = \frac{340 * .04}{30. * \dfrac{\pi}{180}} \approx 26\text{m}$$

One might wonder what the chances are for an aircraft flying at the speed of sound to be within 26 meters from the North Pole. Nevertheless, the argument might continue, it is a possibility and we need to be assured that we can navigate anywhere without encountering infinite ρ_d. Consider a craft that approaches the North Pole then adopting the wander azimuth algorithm will lead to:

As $\phi \cong 90^o \Rightarrow \mathrm{s}(\phi) \cong 1,\ \mathrm{c}(\phi) = \Delta\phi = \dfrac{\pi}{2} - \phi$. Substituting for these values in Eq. (8.3) gives

$$\mathbf{C}_e^n = \begin{bmatrix} \mathrm{c}\alpha\,\Delta\phi & -\mathrm{c}\alpha\,\mathrm{s}\lambda + \mathrm{s}\alpha\,\mathrm{c}\lambda & \mathrm{c}\alpha\,\mathrm{c}\lambda + \mathrm{s}\alpha\,\mathrm{s}\lambda \\ -\mathrm{s}\alpha\,\Delta\phi & \mathrm{s}\alpha\,\mathrm{s}\lambda + \mathrm{c}\alpha\,\mathrm{c}\lambda & -\mathrm{s}\alpha\,\mathrm{c}\lambda + \mathrm{c}\alpha\,\mathrm{s}\lambda \\ -1 & -\Delta\phi\,\mathrm{s}\lambda & \Delta\phi\,\mathrm{c}\lambda \end{bmatrix}$$

Using the trigonometric identities of the sines and cosines of the sum of angles in the above equation gives

$$\mathbf{C}_e^n = \begin{bmatrix} \mathrm{c}\alpha\,\Delta\phi & \mathrm{s}(\alpha-\lambda) & \mathrm{c}(\alpha-\lambda) \\ -\mathrm{s}\alpha\,\Delta\phi & \mathrm{c}(\alpha-\lambda) & -\mathrm{s}(\alpha-\lambda) \\ -1 & -\Delta\phi\,\mathrm{s}\lambda & \Delta\phi\,\mathrm{c}\lambda \end{bmatrix}$$

Finally when substituting for $\Delta\phi$=0 in the above

$$\mathbf{C}_e^n = \begin{bmatrix} 0 & \mathrm{s}(\alpha-\lambda) & \mathrm{c}(\alpha-\lambda) \\ 0 & \mathrm{c}(\alpha-\lambda) & -\mathrm{s}(\alpha-\lambda) \\ -1 & 0 & 0 \end{bmatrix}$$

It can be seen that it is impossible to distinguish between α and λ. This should have been a predictable result! How would it be possible to compute λ with any degree of accuracy when all the longitudes (all real numbers from 0 to 360) meet at one point? This implies that the wander azimuth angle may not be the answer to navigation near the poles. So how can we navigate about the Earth's poles?

8.4 Polar Circle Navigation Algorithm

This algorithm is based on prescribing an arbitrary polar circle centered at the pole with radius, r_{polar} that at its circumference has the latitude ϕ_{polar}. An aircraft will navigate normally using the usual North-East Nav-frame. However, at the instance this craft enters the polar circle, the North East frame is frozen and hence the position DCM, $\mathbf{C}_e^n$, is fixed until the craft leaves the polar circle. With reference to Fig. 5.1, this action is equivalent to forcing ω_{en} to zero. This is essentially the same as assuming the Earth radius of curvature to be infinite or equivalently to plane surface motion. Navigation within this circle will be reduced to two-dimensional Cartesian space. For that, the polar circle must be small enough so that Earth's spherity will not cause significant degradations in the vertical channel.

With reference to Fig. 8.4, we now show how to navigate inside the polar circle and provide the algorithm that will substitute for computing $\mathbf{C}_e^n$. Suppose a craft penetrates the polar circle at the longitude angle, λ_0, and the latitude angle, ϕ_{0polar}. Noting that the craft's horizontal velocity components, v_x and v_y, are along the axes of the frozen frame, the position, p, of the craft inside the circle can then be computed by

$$\Delta x = \int v_x dt \tag{8.9a}$$

$$\Delta y = \int v_y dt \tag{8.9b}$$

$$\Delta\lambda = \tan^{-1}\frac{\Delta y}{r_{polar} - \Delta x} \tag{8.9c}$$

$$\Delta\phi = \frac{\left(r_{polar} - \left|r_{polar} - \Delta x\right|\right)}{r_{polar}}\Delta\phi_{0\,polar} \tag{8.9d}$$

In the above equations, Δx and Δy denote the craft travel along the x- and y-axes inside the polar circle, $\Delta\lambda$ denotes the change in longitude, and $\Delta\phi$ denotes the change in latitude. The longitude and latitude can precisely be determined by the equations

$$\lambda = \lambda_0 + \Delta\lambda$$
$$\phi = \phi_{0\,polar} + \Delta\phi \tag{8.10}$$

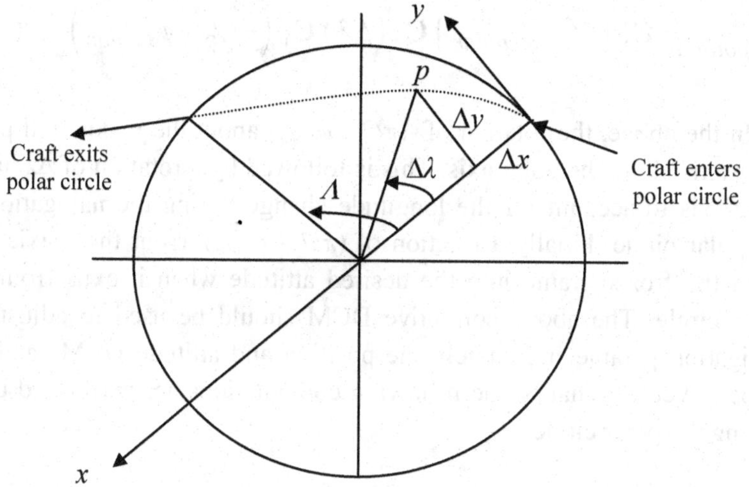

Fig. 8.4a Polar Circle Navigation (Top View)

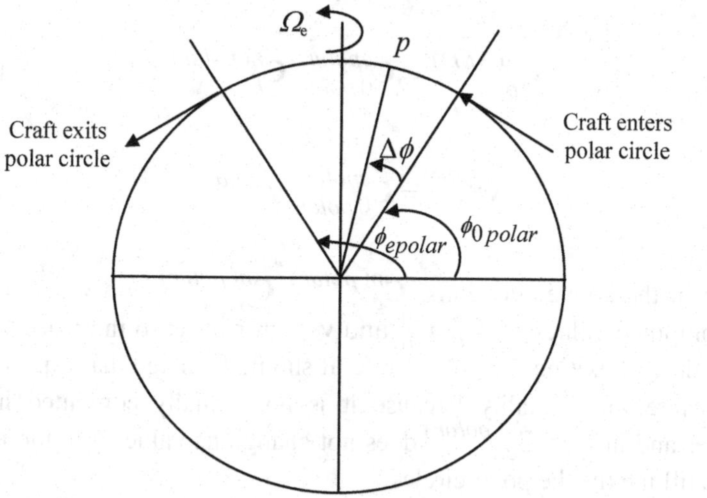

Fig. 8.4b Polar Circle Navigation (Side View)

Now suppose that the craft exits the polar circle when $\Delta\lambda = \lambda_e - \lambda_0$ and the latitude is ϕ_{epolar}, then the change from the (frozen) Nav-frame to the new (exit) Nav-frame is given by this DCM

$$\mathbf{C}_{0\,polar}^{epolar} = \mathbf{C}_y\left(\pi/2 - \phi_{epolar}\right)\mathbf{C}_z\left(\Delta\lambda\right)\mathbf{C}_y\left(-\pi/2 + \phi_{0\,polar}\right) \tag{8.11}$$

In the above, the rotation of $-(\pi/2 - \phi_{0polar})$ about the y-axis will place the z-axis along the polar axis. This is followed by a rotation of $\Delta\lambda$ about the z-axis to account for the longitude change during the navigation in the polar circle. Finally a rotation of $(\pi/2 - \phi_{epolar})$ about the y-axis will bring the frozen frame into the desired attitude when it exits from the polar circle. The above corrective DCM should be used to adjust the navigation parameters, namely the position and attitude DCMs and the velocity vector. That is, the following computations are performed upon exiting the polar circle

$$\mathbf{C}_e^{n(NED)} = \mathbf{C}_{0\,polar}^{epolar}\,\mathbf{C}_e^{n(polar)} \tag{8.12}$$

$$\mathbf{C}_b^{n(NED)} = \mathbf{C}_{0\,polar}^{epolar}\,\mathbf{C}_b^{n(polar)} \tag{8.13}$$

$$\mathbf{v}^{NED} = \mathbf{C}_{0\,polar}^{epolar}\,\mathbf{v}^{polar} \tag{8.14}$$

where in the above equations, $\mathbf{C}_e^{n(polar)}, \mathbf{C}_b^{n(polar)}$ and $\mathbf{v}^{polar}$ denote the computed values of $\mathbf{C}_e^n, \mathbf{C}_b^n$ and $\mathbf{v}$ with respect to the frozen frame at the time of exiting the polar circle. It should be noted that Eq. (8.12) is listed here for formality because it is not actually computed in that manner and in fact $\mathbf{C}_e^{n(polar)}$ does not change its value from the time it enters till it exits the polar circles.

If the quaternion were used in vector transformation, then the following should be performed in lieu of the above equation

$$\mathbf{q}_{0\,polar}^{epolar} = \begin{bmatrix} c\dfrac{\left(\frac{\pi}{2}-\phi_{0\,polar}\right)}{2} \\ 0 \\ -s\dfrac{\left(\frac{\pi}{2}-\phi_{0\,polar}\right)}{2} \\ 0 \end{bmatrix} \begin{bmatrix} c\dfrac{\Delta\lambda}{2} \\ 0 \\ 0 \\ s\dfrac{\Delta\lambda}{2} \end{bmatrix} \begin{bmatrix} c\dfrac{\left(\frac{\pi}{2}-\phi_{epolar}\right)}{2} \\ 0 \\ s\dfrac{\left(\frac{\pi}{2}-\phi_{epolar}\right)}{2} \\ 0 \end{bmatrix} \tag{8.15}$$

The attitude quaternion is corrected as follows

$$\mathbf{q}_b^{n(NED)} = \mathbf{q}_b^{0\,polar}\,\mathbf{q}_{0\,polar}^{epolar} \tag{8.16}$$

where $\mathbf{q}_b^{0\,polar}$ the computed value of $\mathbf{q}_b^n$ with respect to the frozen frame. Similarly, the velocity vector is adjusted to the new NED frame.

We note that the position DCM, $\mathbf{C}_e^n$, is not computed explicitly in the sense that it can always be constructed from the latitude and the longitude values. As a digression, notice that from Eq. (8.2b)

$$\mathbf{C}_e^{n(polar)} = \mathbf{C}_y(-\phi_{0\,polar})\mathbf{C}_x(\lambda_0) \tag{8.17}$$

Substituting from Eqs. (8.11) and (8.17) into Eq. (8.12) results in

$$\begin{aligned}
\mathbf{C}_e^{n(NED)} &= \mathbf{C}_{0\,polar}^{epolar}\,\mathbf{C}_e^{n(polar)} \\
&= \mathbf{C}_y\left(\tfrac{\pi}{2}-\phi_{epolar}\right)\mathbf{C}_z(\Delta\lambda)\mathbf{C}_y\left(-\tfrac{\pi}{2}+\phi_{0\,polar}\right)\mathbf{C}_y\left(-\phi_{0\,polar}\right)\mathbf{C}_x(\lambda_0) \\
&= \mathbf{C}_y\left(-\phi_{epolar}\right)\mathbf{C}_y\left(\tfrac{\pi}{2}\right)\mathbf{C}_z(\Delta\lambda)\mathbf{C}_y\left(-\tfrac{\pi}{2}\right)\mathbf{C}_x(\lambda_0)
\end{aligned}$$

Algebraically, it is straightforward to show that

$$\mathbf{C}_y\left(\pi/2\right)\mathbf{C}_z\left(\Delta\lambda\right)\mathbf{C}_y\left(-\pi/2\right)=\mathbf{C}_x\left(\Delta\lambda\right)$$

Simplifying the above two equation results in the expected expression of

$$\mathbf{C}_e^{n(NED)}=\mathbf{C}_y\left(-\phi_{epolar}\right)\mathbf{C}_x\left(\lambda_0+\Delta\lambda\right)=\mathbf{C}_y\left(-\phi_{epolar}\right)\mathbf{C}_x\left(\lambda_e\right)$$

From a practical point of implementation, it is possible that the navigation algorithm oscillates, in and out of the polar circle, due to navigation errors. To alleviate the computational annoyance due to these oscillations, we prescribe a computational 'relay' that includes two concentric rings.

The polar navigation is switched on when a craft crosses the inner circle and is switched off only when the craft leaves the outer circle as shown in Fig.8.5.

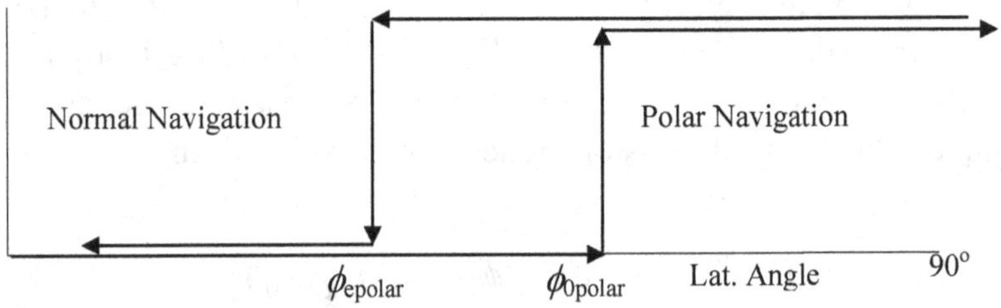

Fig. 8.5 Logic for Implementing Polar Navigation Solution

8.5 Alternative Polar Circle Navigation Frame

In the above algorithm the navigation frame has been frozen on the rim of the polar circle to avoid the computational problems. Alternatively we could apply another algorithm in which the Nav-frame moves with the body inside the polar circle but is not allowed to rotate about the *z*-axis. This is almost equivalent to the wander azimuth algorithm but without the associated computational burden. In this algorithm, the Earth rate will remain the same as outside the polar circle as

$$\boldsymbol{\omega}_{ie}^{n} = \Omega_e \begin{bmatrix} c\phi \\ 0 \\ -s\phi \end{bmatrix}$$

With reference to Fig. 5.1, the craft rate will be computed by

$$\boldsymbol{\omega}_{en}^{n} = \begin{bmatrix} \dot{\lambda}\cos(\phi) \\ -\dot{\phi} \\ 0 \end{bmatrix} = \begin{bmatrix} \dfrac{v_e}{R_p + h} \\ -\dfrac{v_n}{R_m + h} \\ 0 \end{bmatrix} \tag{8.18}$$

Notice that inside the polar circle, discontinuity in the axes directions is not allowed, and hence the latitude magnitude should always increase. For example, if a craft flies head on to the pole, the latitude, ϕ, will increase until it reaches 90° on the pole and when it passes the pole, ϕ should continue to increase past 90°.

Similar to the fixed frame algorithm, the DCM

$$\mathbf{C}_{0\,polar}^{epolar} = \mathbf{C}_z\left(\Delta\lambda\right) \tag{8.19}$$

is used to correct for both the attitude DCM and for the velocity, the same as in Eqs. (8.13) and (8.14). If the quaternions were used, then the following should be performed in lieu of the above equation

$$\mathbf{q}_{0polar}^{epolar} = \begin{bmatrix} c\frac{\Delta\lambda}{2} \\ 0 \\ 0 \\ s\frac{\Delta\lambda}{2} \end{bmatrix} \tag{8.20}$$

In the following table we list the differences in computations with each choice.

Table 8.1 Computations in the Alternative Algorithms inside the Polar Circle

	Moving Frame	Fixed Frame
Earth Rate	$\boldsymbol{\omega}_{ie}^{n} = \Omega_e \begin{bmatrix} c\phi \\ 0 \\ -s\phi \end{bmatrix}$ ϕ must be current and increasing past the pole	$\boldsymbol{\omega}_{ie}^{n} = \Omega_e \begin{bmatrix} c\phi_{0polar} \\ 0 \\ -s\phi_{0polar} \end{bmatrix}$ ϕ_{0polar} is the latitude at the rim of the polar circle
Craft Rate	$\boldsymbol{\omega}_{en}^{n} = \begin{bmatrix} \dot{\lambda} c\phi \\ -\dot{\phi} \\ 0 \end{bmatrix}$	$\boldsymbol{\omega}_{en}^{n} = \begin{bmatrix} 0 \\ 0 \\ 0 \end{bmatrix}$
Exiting Polar Circle	Rotate $\mathbf{v}_{en}^{n}$ and $\mathbf{C}_b^n$ by $\Delta\lambda$ about the z-axis.	Rotate $\mathbf{v}_{en}^{n}$ and $\mathbf{C}_b^n$ by $\mathbf{C}_{0polar}^{epolar}$.

From the above table, it appears that the moving frame may be a better choice because of the last condition. From an implementation viewpoint, it is less work to compute the corrective transformation using Eq. (8.19) than Eq. (8.11).

The past few chapters should enable us to solve the navigation equation anywhere. What remains is to initialize the navigation parameters. Like any set of differential equations, initial conditions should be determined to solve it. This is addressed in the following chapter.

References

Boeing Aero, "Polar Route Operations," http://www.boeing.com/commercial/aeromagazine/aero_16/polar.html

Chapter 9

Alignment

9.1 Introduction

By now we know how to implement and solve for the differential equations that compute the navigation equations parameters. What remains – like for solving any differential equation – is initializing these parameters. Traditionally this process is called alignment, a term inherited from the legacy gimbaled systems because it mechanically aligns the stable platform to the navigation frame. Here we have a body-fixed inertial suite that is oriented in some arbitrary direction, not the mechanical frame. To start the solution we need to determine the craft initial attitude and heading as well as its initial position and velocity in the navigation frame. Ideally, we could use the inertial sensor outputs to execute the alignment process. However, we are facing the challenge that the accelerometer and gyro are contaminated with many error sources [1-3]. Even though these data are preprocessed to account for known error sources, the process is not exhaustive. Other error sources are still unaccounted for because some of these errors are random and can't be calibrated against. Nevertheless, some errors can be estimated, chief amongst them are the accelerometer and gyro residual biases that varies from one power-on to another.

The craft's position coordinates– longitude, latitude and altitude – are initialized manually or by reading stored data from its past trajectory. Its velocity components are initialized with zeros if it is grounded or from an external data source such as GPS if in motion. This leaves us with the craft attitude, accelerometer bias, and gyro drift to initialize. Initializing these elements depends on the available sources of data and their associated technology. For example, highly accurate inertial sensors – typically classified as navigation grade sensors because they are capable of performing the navigation solution – could be relied on for performing the alignment as is discussed herein. Conversely, as in the emerging micro electro-mechanical sensors (MEMS) – classified as tactical grade sensors – the gyro errors could be so high that the outputs are rendered useless for performing the alignment. This is discussed in the following chapter.

9.2 IMU Alignment

Here both the accelerometer and gyro sets are used for alignment. Initially it will be assumed that their measurements are perfect. That is, both the gyros and accelerometers are mutually orthogonal, are aligned perfectly to the principal body axes of the craft, and are not corrupted by any sources of errors.

Let us address first the attitude initialization and recall that attitude information is embodied in the DCM $\mathbf{C}_b^n$. One would expect that the sensed accelerometer vector, when transformed into the navigation frame, to align perfectly with the gravity specific force, i.e.

$$\mathbf{C}_b^n \mathbf{a}_s = -\mathbf{g}^n = -g \begin{pmatrix} 0 \\ 0 \\ 1 \end{pmatrix} \tag{9.1}$$

where $\mathbf{a}_s$ is the sensed accelerometer vector. Likewise, we should see that the gyro sensor vector, when transformed into the navigation frame, aligns perfectly with the Earth rotation rate, i.e.

$$\mathbf{C}_b^n \omega_s = \mathbf{\Omega}^n = \Omega_e \begin{pmatrix} \cos\phi \\ 0 \\ -\sin\phi \end{pmatrix} \tag{9.2}$$

where $\boldsymbol{\omega}_s$ is the sensed gyro vector and ϕ is the latitude of the craft. One way to determine $\mathbf{C}_b^n$, as given in Britting [4], is to solve for the following equation

$$\mathbf{C}_b^n\left(-\mathbf{a}_s \quad \boldsymbol{\omega}_s \quad -\mathbf{a}_s\times\boldsymbol{\omega}_s\right)=\left(\mathbf{g}^n \quad \boldsymbol{\Omega}^n \quad \mathbf{g}^n\times\boldsymbol{\Omega}^n\right) \tag{9.3}$$

With little algebraic manipulation, the solution of the above equation is given by

$$\mathbf{C}_n^b=\left(\boldsymbol{\omega}_s \quad -\mathbf{a}_s \quad -\mathbf{a}_s\times\boldsymbol{\omega}_s\right)\left(\boldsymbol{\Omega}^n \quad \mathbf{g}^n \quad \mathbf{g}^n\times\boldsymbol{\Omega}^n\right)^{-1} \tag{9.4}$$

Later, Jiang et al [5] modified the solution to

$$\mathbf{C}_n^b=\left(-\mathbf{a}_s \quad -\mathbf{a}_s\times\boldsymbol{\omega}_s \quad (\mathbf{a}_s\times\boldsymbol{\omega}_s)\times\mathbf{a}_s\right)\left(\mathbf{g}^n \quad \mathbf{g}^n\times\boldsymbol{\Omega}^n \quad (\mathbf{g}^n\times\boldsymbol{\Omega}^n)\times\mathbf{g}^n\right)^{-1} \tag{9.5}$$

Since in reality both the gyro and accelerometer measurements are erroneous, neither equation (9.4) nor (9.5) will provide an authentic DCM, (i.e. a unitary matrix).

We now improve on Eqs. (9.4) and (9.5) and introduce a procedure that leads to a unitary DCM and provides optimal estimates of the accelerometer and gyro biases. The method is based on orthonormalizing the matrices in the above equation. To do that let us permute the columns in Eq. (9.5) to

$$\mathbf{C}_n^b=\left((\mathbf{a}_s\times\boldsymbol{\omega}_s)\times\mathbf{a}_s \quad -\mathbf{a}_s\times\boldsymbol{\omega}_s \quad -\mathbf{a}_s\right)\left((\mathbf{g}^n\times\boldsymbol{\Omega}^n)\times\mathbf{g}^n \quad \mathbf{g}^n\times\boldsymbol{\Omega}^n \quad \mathbf{g}^n\right)^{-1} \tag{9.6}$$

These changes will cause the columns in each matrix to appear in a vector right hand system. Applying the vector cross product rule to the second column of the RHS matrix, referring to Fig. 9.1 for angles, yields

$$\left|\mathbf{g}^n \times \mathbf{\Omega}^n\right| = \left|\mathbf{g}^n\right| \cdot \left|\mathbf{\Omega}^n\right| \cdot \sin(\mathbf{g}^n, \mathbf{\Omega}^n) = g\Omega_e \sin\left(\frac{\pi}{2} + \phi\right) = g\Omega_e \cos(\phi)$$

where ϕ is the true geodetic latitude. Hence

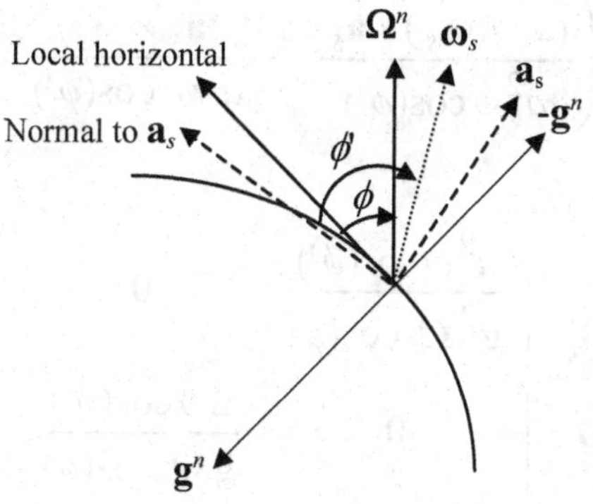

Figure 9.1 Physical and Measured Vectors

$$\begin{pmatrix}(\mathbf{g}^n \times \mathbf{\Omega}^n) \times \mathbf{g}^n & \mathbf{g}^n \times \mathbf{\Omega}^n & \mathbf{g}^n\end{pmatrix}$$

$$= \begin{pmatrix}\dfrac{(\mathbf{g}^n \times \mathbf{\Omega}^n) \times \mathbf{g}^n}{g^2\Omega_e \cos(\phi)} & \dfrac{\mathbf{g}^n \times \mathbf{\Omega}^n}{g\Omega_e \cos(\phi)} & \dfrac{\mathbf{g}^n}{g}\end{pmatrix} \begin{bmatrix} g^2\Omega_e \cos(\phi) & 0 & 0 \\ 0 & g\Omega_e \cos(\phi) & 0 \\ 0 & 0 & g \end{bmatrix} \tag{9.7}$$

Similarly

$$\left((\mathbf{a}_S \times \boldsymbol{\omega}_S) \times \mathbf{a}_S \quad -\mathbf{a}_S \times \boldsymbol{\omega}_S \quad -\mathbf{a}_S \right)$$

$$= \left(\frac{(\mathbf{a}_S \times \boldsymbol{\omega}_S) \times \mathbf{a}_S}{a^2 \omega \cos(\phi')} \quad -\frac{\mathbf{a}_S \times \boldsymbol{\omega}_S}{a\omega \cos(\phi')} \quad -\frac{\mathbf{a}_S}{a} \right) \begin{pmatrix} a^2 \omega \cos(\phi') & 0 & 0 \\ 0 & a\omega \cos(\phi') & 0 \\ 0 & 0 & a \end{pmatrix}$$

(9.8)

where in the above equation, $a = |\mathbf{a}_S|$, $\omega = |\boldsymbol{\omega}_S|$ and ϕ' is the complement of the angle between $\mathbf{a}_S$ and $\boldsymbol{\omega}_S$. Let

$$\mathbf{E} = \left(\frac{(\mathbf{a}_S \times \boldsymbol{\omega}_S) \times \mathbf{a}_S}{a^2 \omega \cos(\phi')} \quad -\frac{\mathbf{a}_S \times \boldsymbol{\omega}_S}{a\ \omega\ \cos(\phi')} \quad -\frac{\mathbf{a}_S}{a} \right) \tag{9.9a}$$

$$\mathbf{D} = \begin{pmatrix} \frac{a^2 \omega \cos(\phi')}{g^2 \Omega_e \cos(\phi)} & 0 & 0 \\ 0 & \frac{a\omega \cos(\phi')}{g \Omega_e \cos(\phi)} & 0 \\ 0 & 0 & \frac{a}{g} \end{pmatrix} \tag{9.9b}$$

$$\mathbf{F} = \left(\frac{(\mathbf{g}^n \times \boldsymbol{\Omega}^n) \times \mathbf{g}^n}{g^2 \Omega_e \cos(\phi)} \quad \frac{\mathbf{g}^n \times \boldsymbol{\Omega}^n}{g \Omega_e \cos(\phi)} \quad \frac{\mathbf{g}^n}{g} \right) \tag{9.9c}$$

Observe that the matrices **E** and **F** are orthonormal by construction; hence they are legitimate DCMs. Substituting from Eqs. (9.7)-(9.9) into (9.6) yields

$$\mathbf{C}_n^b = \mathbf{E}\,\mathbf{D}\,\mathbf{F}^{-1} = \mathbf{E}\,\mathbf{D}\,\mathbf{F}' \tag{9.10}$$

Notice also that potential errors in the **D** matrix could result only from biases or scale factors. Since these terms are not known upfront, then intuitively, the best available estimate to $\mathbf{C}_n^b$ is

$$\hat{\mathbf{C}}_n^b = \mathbf{E}\,\mathbf{F}' \tag{9.11}$$

Algebraically, Eq. (9.10) could be represented as the product of two matrices

$$\mathbf{C}_n^b = \mathbf{Q}\mathbf{S} \tag{9.12}$$

where $\mathbf{Q} = \mathbf{E}\mathbf{F}'$ is an orthonormal matrix and $\mathbf{S} = \mathbf{F}\,\mathbf{D}\,\mathbf{F}'$ is a symmetric matrix. This way, **Q** is the matrix responsible for rotation and **S** is the matrix responsible for distortion [6]. Hence an orthonormalized $\mathbf{C}_n^b$ is given by

$$\hat{\mathbf{C}}_n^b = \mathbf{Q} = \mathbf{E}\,\mathbf{F}' \tag{9.13}$$

Therefore $\mathbf{C}_n^b$ orthonormalized by either Eq. (9.11) or Eq. (9.13) is the same. Essentially, this tells us that errors due to biases or scale factors do not contribute to the estimation of an orthonormalized $\mathbf{C}_n^b$. Of interest is to determine the symmetric matrix **S**. By multiplying each side in Eq. (9.12) by its transpose we get

$$(\mathbf{C}_n^b)'\mathbf{C}_n^b = \mathbf{S}'\mathbf{Q}'\mathbf{Q}\mathbf{S} = \mathbf{S}\mathbf{S} = \mathbf{S}^2$$

which implies that

$$\mathbf{S} = \left((\mathbf{C}_n^b)'\mathbf{C}_n^b\right)^{\frac{1}{2}}$$

Right multiplying both sides of Eq. (9.12) by $\mathbf{S}^{-1}$

$$\mathbf{Q} = \mathbf{C}_n^b \left((\mathbf{C}_n^b)' \mathbf{C}_n^b \right)^{-\frac{1}{2}}$$

one would have arrived at the same result as in [4,5]. We now cast $\mathbf{C}_n^b$ in a more convenient form. Notice that from Eq. (9.1)

$$\mathbf{g}^n = g \begin{pmatrix} 0 \\ 0 \\ 1 \end{pmatrix} \tag{9.14a}$$

Using Eq. (9.2) we get

$$\frac{\mathbf{g}^n \times \mathbf{\Omega}^n}{g \Omega_e \cos\phi} = \frac{1}{g \Omega_e \cos\phi} \begin{pmatrix} 0 \\ 0 \\ g \end{pmatrix} \times \Omega_e \begin{pmatrix} \cos\phi \\ 0 \\ -\sin\phi \end{pmatrix} = \begin{pmatrix} 0 \\ 1 \\ 0 \end{pmatrix} \tag{9.14b}$$

Finally from the above two equations we find that

$$\frac{(\mathbf{g}^n \times \mathbf{\Omega}^n) \times \mathbf{g}^n}{g^2 \Omega_e \cos\phi} = \begin{pmatrix} 1 \\ 0 \\ 0 \end{pmatrix} \tag{9.14c}$$

Now from Eqs. (9.9c) and (9.14) we can find that

$$\mathbf{F} = \begin{pmatrix} \dfrac{(\mathbf{g}^n \times \mathbf{\Omega}^n) \times \mathbf{g}^n}{g^2 \Omega_e \cos\phi} & \dfrac{\mathbf{g}^n \times \mathbf{\Omega}^n}{g \Omega_e \cos\phi} & \dfrac{\mathbf{g}^n}{g} \end{pmatrix} = \begin{pmatrix} 1 & 0 & 0 \\ 0 & 1 & 0 \\ 0 & 0 & 1 \end{pmatrix} = \mathbf{I} \tag{9.15}$$

Therefore

$$\mathbf{C}_n^b = \mathbf{E} = \begin{pmatrix} \dfrac{(\mathbf{a}_s \times \boldsymbol{\omega}_s) \times \mathbf{a}_s}{a^2 \omega \cos\phi'} & -\dfrac{\mathbf{a}_s \times \boldsymbol{\omega}_s}{a\omega \cos\phi'} & -\dfrac{\mathbf{a}_s}{a} \end{pmatrix} \tag{9.16}$$

The above equation could be further simplified. Let

$$\begin{aligned} \bar{\mathbf{a}} &= \frac{\mathbf{a}_s}{a} \\ \bar{\boldsymbol{\omega}} &= \frac{\boldsymbol{\omega}_s}{\omega} \end{aligned} \tag{9.17}$$

then Eq. (9.16) becomes

$$\mathbf{C}_n^b = \begin{pmatrix} \dfrac{(\bar{\mathbf{a}} \times \bar{\boldsymbol{\omega}}) \times \bar{\mathbf{a}}}{\cos\phi'} & -\dfrac{\bar{\mathbf{a}} \times \bar{\boldsymbol{\omega}}}{\cos\phi'} & -\bar{\mathbf{a}} \end{pmatrix} \tag{9.18}$$

Expanding the cross product in the above yields

$$\mathbf{C}_n^b = \begin{pmatrix} \dfrac{\bar{\boldsymbol{\omega}} - \sin\phi' \bar{\mathbf{a}}}{\cos\phi'} & -\dfrac{\bar{\mathbf{a}} \times \bar{\boldsymbol{\omega}}}{\cos\phi'} & -\bar{\mathbf{a}} \end{pmatrix} \tag{9.19}$$

Note that from a physical standpoint, Eq. (9.18) behaves like a procedure for generating three orthonormalized vectors from the two IMU sensors, $\mathbf{a}_s$ and $\boldsymbol{\omega}_s$. First, we used the accelerometer sensor unit vector, $\bar{\mathbf{a}}$, as a reference vector to be the third column of $\mathbf{C}_n^b$, (z-axis).

Next, the vector cross product $(\mathbf{a}_s \times \boldsymbol{\omega}_s)$ of the sensor's two unit vectors is orthogonal to their co plane – ideally the azimuth plane – and becomes the second column, (y-axis). Finally, the cross product of the second and third columns is in the azimuth plane and orthogonal to the z-axis, and forms the first column, (x-axis). The process of construction of

$\mathbf{C}_n^b$ is shown graphically in Figure 9.2, where $\mathbf{a}_s$ is used as the lead z-axis vector.

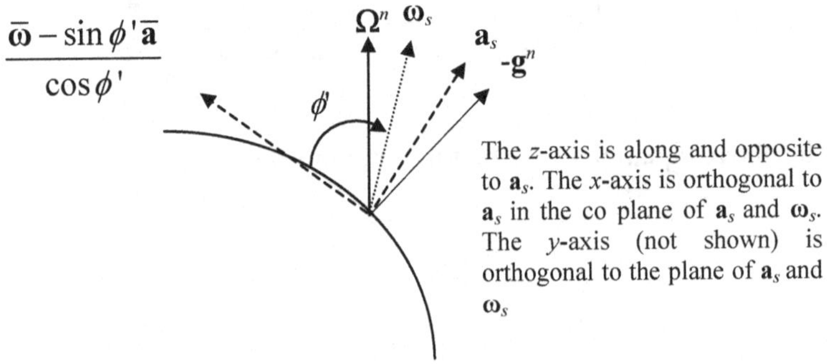

Figure 9.2 Construction of $\mathbf{C}_b^n$, $\mathbf{a}_s$ is the lead vector

9.3 Alternative Algorithm for $\mathbf{C}_n^b$

In above algorithm the accelerometer vector was selected to be the lead vector for constructing an orthonormalized set. This arouses some curiosity about what we would have gotten had we used the Earth rotation vector as the reference vector instead of the gravity vector. This is explored herein.

Following the steps that led to the construction of Eq. (9.6) we can write the new $\mathbf{C}_b^n$ as

$$\mathbf{C}_n^b = \begin{pmatrix} \boldsymbol{\omega}_s & -\mathbf{a}_s \times \boldsymbol{\omega}_s & -\boldsymbol{\omega}_s \times (\mathbf{a}_s \times \boldsymbol{\omega}_s) \end{pmatrix} \begin{pmatrix} \boldsymbol{\Omega}^n & \mathbf{g}^n \times \boldsymbol{\Omega}^n & \boldsymbol{\Omega}^n \times (\mathbf{g}^n \times \boldsymbol{\Omega}^n) \end{pmatrix}^{-1} \tag{9.20}$$

The matrices in the above equation can be orthogonalized as follows

$$\begin{pmatrix}\boldsymbol{\Omega}^n & \mathbf{g}^n\times\boldsymbol{\Omega}^n & \boldsymbol{\Omega}^n\times(\mathbf{g}^n\times\boldsymbol{\Omega}^n)\end{pmatrix}$$
$$=\begin{pmatrix}\dfrac{\boldsymbol{\Omega}^n}{\Omega_e} & \dfrac{\mathbf{g}^n\times\boldsymbol{\Omega}^n}{g\Omega_e\cos\phi} & \dfrac{\boldsymbol{\Omega}^n\times(\mathbf{g}^n\times\boldsymbol{\Omega}^n)}{g\Omega_e^2\cos\phi}\end{pmatrix}\begin{pmatrix}\Omega_e & 0 & 0\\ 0 & g\Omega_e\cos\phi & 0\\ 0 & 0 & g\Omega_e^2\cos\phi\end{pmatrix} \tag{9.21}$$

Similarly

$$\begin{pmatrix}\boldsymbol{\omega}_s & -\mathbf{a}_s\times\boldsymbol{\omega}_s & -\boldsymbol{\omega}_s\times(\mathbf{a}_s\times\boldsymbol{\omega}_s)\end{pmatrix}$$
$$=\begin{pmatrix}\dfrac{\boldsymbol{\omega}_s}{\omega} & -\dfrac{\mathbf{a}_s\times\boldsymbol{\omega}_s}{a\omega\cos\phi'} & -\dfrac{\boldsymbol{\omega}_s\times(\mathbf{a}_s\times\boldsymbol{\omega}_s)}{a\omega^2\cos\phi'}\end{pmatrix}\begin{pmatrix}\omega & 0 & 0\\ 0 & a\omega\cos\phi' & 0\\ 0 & 0 & a\omega^2\cos\phi'\end{pmatrix} \tag{9.22}$$

Let **E**, **D**, **F** be as follows

$$\mathbf{E}=\begin{pmatrix}\dfrac{\boldsymbol{\omega}_s}{\omega} & -\dfrac{\mathbf{a}_s\times\boldsymbol{\omega}_s}{a\omega\cos\phi'} & -\dfrac{\boldsymbol{\omega}_s\times(\mathbf{a}_s\times\boldsymbol{\omega}_s)}{a\omega^2\cos\phi'}\end{pmatrix}$$

Substituting from Eq. (9.17)

$$\begin{aligned}\mathbf{E}&=\begin{pmatrix}\bar{\boldsymbol{\omega}} & -\dfrac{\bar{\mathbf{a}}\times\bar{\boldsymbol{\omega}}}{\cos\phi'} & -\dfrac{\bar{\boldsymbol{\omega}}\times(\bar{\mathbf{a}}\times\bar{\boldsymbol{\omega}})}{\cos\phi'}\end{pmatrix}\\ &=\begin{pmatrix}\bar{\boldsymbol{\omega}} & -\dfrac{\bar{\mathbf{a}}\times\bar{\boldsymbol{\omega}}}{\cos\phi'} & -\dfrac{\bar{\mathbf{a}}-\sin(\phi')\bar{\boldsymbol{\omega}})}{\cos\phi'}\end{pmatrix}\end{aligned} \tag{9.23}$$

$$\mathbf{F}=\begin{pmatrix}\dfrac{\boldsymbol{\Omega}^n}{\Omega_e} & \dfrac{\mathbf{g}^n\times\boldsymbol{\Omega}^n}{g\Omega_e\cos\phi} & \dfrac{\boldsymbol{\Omega}^n\times(\mathbf{g}^n\times\boldsymbol{\Omega}^n)}{g\Omega_e^2\cos\phi}\end{pmatrix} \tag{9.24}$$

$$\mathbf{D} = \begin{pmatrix} \frac{\omega}{\Omega_e} & 0 & 0 \\ 0 & \frac{a\omega \cos\phi'}{g\Omega_e \cos\phi} & 0 \\ 0 & 0 & \frac{a\omega^2 \cos\phi'}{g\Omega_e^2 \cos\phi} \end{pmatrix} \tag{9.25}$$

As before

$$\mathbf{C}_n^b = \mathbf{E}\,\mathbf{D}\,\mathbf{F}' \tag{9.26}$$

Orthogonalizing $\mathbf{C}_n^b$ yields

$$\mathbf{C}_n^b = \mathbf{E}\,\mathbf{F}' \tag{9.27}$$

Since

$$\mathbf{F} = \begin{pmatrix} \frac{\mathbf{\Omega}^n}{\Omega_e} & \frac{\mathbf{g}^n \times \mathbf{\Omega}^n}{g\Omega_e \cos\phi} & \frac{\mathbf{\Omega}^n \times (\mathbf{g}^n \times \mathbf{\Omega}^n)}{g\Omega_e^2 \cos\phi} \end{pmatrix} = \begin{pmatrix} \cos\phi & 0 & \sin\phi \\ 0 & 1 & 0 \\ -\sin\phi & 0 & \cos\phi \end{pmatrix} \tag{9.28}$$

Substituting for **E** and **F** from Eqs. (9.23) and (9.28) respectively in Eq. (9.27) gives (using the short hand notations for the sines and cosines)

$$\mathbf{C}_n^b = \begin{pmatrix} \overline{\boldsymbol{\omega}} & -\frac{\overline{\mathbf{a}} \times \overline{\boldsymbol{\omega}}}{c\phi'} & -\frac{\overline{\mathbf{a}} - s\phi'\,\overline{\boldsymbol{\omega}})}{c\phi'} \end{pmatrix} \begin{pmatrix} c\phi & 0 & -s\phi \\ 0 & 1 & 0 \\ s\phi & 0 & c\phi \end{pmatrix}$$
$$= \begin{pmatrix} c\phi\,\overline{\boldsymbol{\omega}} - s\phi\frac{\overline{\mathbf{a}} - s\phi'\overline{\boldsymbol{\omega}}}{c\phi'} & -\frac{\overline{\mathbf{a}} \times \overline{\boldsymbol{\omega}}}{c\phi'} & -s\phi\,\overline{\boldsymbol{\omega}} - c\phi\frac{\overline{\mathbf{a}} - s\phi'\overline{\boldsymbol{\omega}})}{c\phi'} \end{pmatrix} \tag{9.29}$$

Using the trigonometric laws for the sine and cosine of the sum of angles in the above gives

$$\mathbf{C}_n^b = \begin{pmatrix} \dfrac{\cos(\phi' - \phi)\bar{\boldsymbol{\omega}} - \sin(\phi)\bar{\mathbf{a}}}{\cos\phi'} & -\dfrac{\bar{\mathbf{a}} \times \bar{\boldsymbol{\omega}}}{\cos\phi'} & \dfrac{\sin(\phi' - \phi)\bar{\boldsymbol{\omega}} - \cos(\phi)\bar{\mathbf{a}}}{\cos\phi'} \end{pmatrix} \quad (9.30)$$

Figure 9.3 depicts the construction of $\mathbf{C}_n^b$.

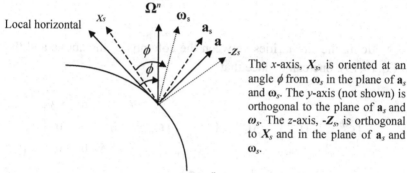

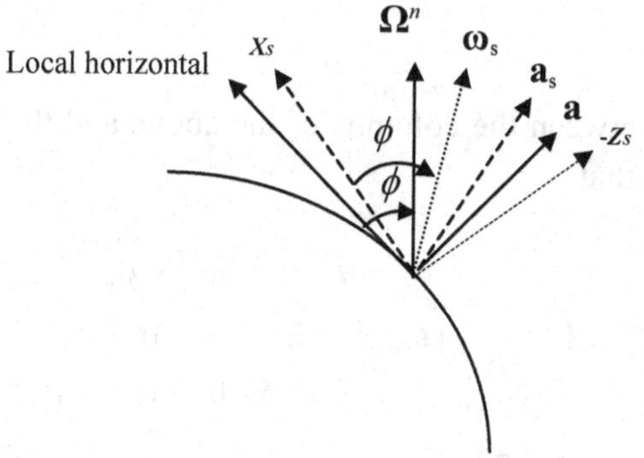

The x-axis, X_s, is oriented at an angle ϕ from $\boldsymbol{\omega}_s$ in the plane of $\mathbf{a}_s$ and $\boldsymbol{\omega}_s$. The y-axis (not shown) is orthogonal to the plane of $\mathbf{a}_s$ and $\boldsymbol{\omega}_s$. The z-axis, $-Z_s$, is orthogonal to X_s and in the plane of $\mathbf{a}_s$ and $\boldsymbol{\omega}_s$.

Figure 9.3 Construction of $\mathbf{C}_b^n$, $\boldsymbol{\omega}_s$ is the lead vector

We now discuss the difference between Eq. (9.30) and its counterpart given by Eq. (9.19). In both equations the second column is the same. This column, as mentioned earlier, is perpendicular to the plane of the IMU unit vectors, and thus is not affected by the choice of the bases. It is noted also that if $\phi' = \phi$ these two equations will again be exactly the same.

Let $\left(\mathbf{C}_n^b\right)_1$ and $\left(\mathbf{C}_n^b\right)_2$ be the DCM's given in Eqs. (9.19) and (9.30), respectively, then the difference between them is

$$\Delta\mathbf{C}_n^b = \left(\mathbf{C}_n^b\right)_2 - \left(\mathbf{C}_n^b\right)_1$$
$$= \frac{1}{\cos(\phi')}\left[\begin{pmatrix} [\cos(\phi' - \phi) - 1]\bar{\boldsymbol{\omega}} \\ +[\sin(\phi') - \sin(\phi)]\bar{\mathbf{a}} \end{pmatrix} \quad \mathbf{0} \quad \begin{pmatrix} \sin(\phi' - \phi)\bar{\boldsymbol{\omega}} \\ +[\cos(\phi') - \cos(\phi)]\bar{\mathbf{a}} \end{pmatrix}\right]$$

(9.31)

Assuming the difference $\Delta\phi = \phi' - \phi$ is small, expanding the trigonometric terms with a Taylor series and preserving first order error terms Eq. (9.31) becomes

$$\Delta \mathbf{C}_n^b = \begin{pmatrix} \Delta\phi\,\bar{\mathbf{a}} & \mathbf{0} & \dfrac{\Delta\phi\,\bar{\boldsymbol{\omega}} - \sin(\phi')\Delta\phi\,\bar{\mathbf{a}}}{\cos(\phi')} \end{pmatrix}$$
$$= \Delta\phi \begin{pmatrix} \bar{\mathbf{a}} & \mathbf{0} & \dfrac{\bar{\boldsymbol{\omega}} - \sin(\phi')\,\bar{\mathbf{a}}}{\cos(\phi')} \end{pmatrix} \qquad (9.32)$$

Noticing the similarities between the columns of the above and those in Eq. (9.19), one can deduce that

$$\left(\mathbf{C}_n^b\right)_2 = \left(\mathbf{C}_n^b\right)_1 + \Delta\mathbf{C}_n^b = \left(\mathbf{C}_n^b\right)_1 + \left(\mathbf{C}_n^b\right)_1 \begin{pmatrix} 0 & 0 & \Delta\phi \\ 0 & 0 & 0 \\ -\Delta\phi & 0 & 0 \end{pmatrix} \qquad (9.33)$$
$$= \left(\mathbf{C}_n^b\right)_1 \begin{pmatrix} 1 & 0 & \Delta\phi \\ 0 & 1 & 0 \\ -\Delta\phi & 0 & 1 \end{pmatrix}$$

which implies that $\left(\mathbf{C}_\text{n}^\text{b}\right)_2$ is just a rotation of $\left(\mathbf{C}_\text{n}^\text{b}\right)_1$ about the y-axis by an angle $\Delta\phi$. Therefore, if $\phi' = \phi$ then $\mathbf{C}_n^b$ in both Eqs. (9.19) and (9.30) will be identical as shown earlier. The above analysis shows the alternatives for selecting $\mathbf{C}_n^b$:

1. Use Eq. (9.19)
2. Use Eq. (9.30) or
3. Use a combination of both.

The last choice is handled mathematically by selecting a fraction, *m*, between 0 and 1 and rotating $\mathbf{C}_n^b$ in Eq. (9.18) by an angle of $m\Delta\phi$ about the *y*-axis, and could be used as a guideline. For example, if we know that the accelerometer data are more reliable, then they should have higher weight than the gyro's, and *m* should be close to zero (and vice versa).

9.4 Estimation of the Accelerometer and Gyro Biases

Ideally, the bias in a certain measurement is the difference between the measured and the nominal values. As such, the acceleration bias, $\mathbf{b}_{accel}$, is given by

$$\begin{aligned}\mathbf{b}_{accel} &= \mathbf{a}_s - \mathbf{C}_n^b \mathbf{a}^n_{nominal} \\ &= \mathbf{a}_s - \mathbf{C}_n^b(-\mathbf{g}^n)\end{aligned} \tag{9.34}$$

Substituting from Eq. (9.1) for $\mathbf{g}^n$ we get

$$\begin{aligned}\mathbf{b}_{accel} &= \mathbf{a}_s + g\mathbf{C}_n^b \begin{pmatrix} 0 \\ 0 \\ 1 \end{pmatrix} \\ &= a\bar{\mathbf{a}} + g(-\bar{\mathbf{a}}) \\ &= (a-g)\bar{\mathbf{a}}\end{aligned} \tag{9.35}$$

The second equality results from Eqs. (9.16) and (9.17). Similarly the gyro drift, $\mathbf{d}_{gyro}$, is given by

$$\mathbf{d}_{gyro} = \boldsymbol{\omega}_s - \mathbf{C}_n^b \boldsymbol{\Omega}^n \tag{9.36}$$

Substituting from Eq. (9.2) for $\boldsymbol{\Omega}^n$ gives

$$\mathbf{d}_{gyro} = \boldsymbol{\omega}_s - \Omega_e \mathbf{C}_n^b \begin{pmatrix} \cos\phi \\ 0 \\ -\sin\phi \end{pmatrix}$$

Substituting for $\mathbf{C}_n^b$ from Eq. (9.19) in the above equation and simplifying gives

$$\mathbf{d}_{gyro} = \boldsymbol{\omega}_s - \Omega_e\left(\cos(\phi)\frac{\bar{\boldsymbol{\omega}} - \sin(\phi')\bar{\mathbf{a}}}{\cos(\phi')} + \sin(\phi)\bar{\mathbf{a}}\right)$$
$$= \left(\omega - \Omega_e\frac{\cos(\phi)}{\cos(\phi')}\right)\bar{\boldsymbol{\omega}} + \Omega_e\frac{\sin(\phi-\phi')}{\cos(\phi')}\bar{\mathbf{a}} \tag{9.37}$$

9.5 Effects of Biases on Estimate of $\mathbf{C}_n^b$

What would be the estimate of $\mathbf{C}_n^b$ if the above biases were incorporated in the measured signals of the sensors? As shown in Eq. (9.6), $\mathbf{C}_n^b$ is the product of two matrices, the first is a function of the sensor measurements and the other is just a constant matrix. From Eqs. (9.35) and (9.36), the adjusted accelerometer and gyro (when the biases are accounting for) become

$$\hat{\mathbf{a}} = \mathbf{a}_s - \mathbf{b}_{accel} = \mathbf{a} - \mathbf{a} - \mathbf{C}_n^b\mathbf{g}^n = -\mathbf{C}_n^b\mathbf{g}^n$$
$$\hat{\boldsymbol{\omega}} = \boldsymbol{\omega}_s - \mathbf{d}_{gyro} = \boldsymbol{\omega}_s - \boldsymbol{\omega}_s + \mathbf{C}_n^b\boldsymbol{\Omega}^n = \mathbf{C}_n^b\boldsymbol{\Omega}^n \tag{9.38}$$

From the above we see that

$$\hat{\mathbf{a}}\times\hat{\boldsymbol{\omega}} = -\left(\mathbf{C}_n^b\mathbf{g}^n\right)\times\left(\mathbf{C}_n^b\boldsymbol{\Omega}^n\right) = -\mathbf{C}_n^b\left(\mathbf{g}^n\times\boldsymbol{\Omega}^n\right)$$
$$\left(\hat{\mathbf{a}}\times\hat{\boldsymbol{\omega}}\right)\times\hat{\mathbf{a}} = \left(-\mathbf{C}_n^b\left(\mathbf{g}^n\times\boldsymbol{\Omega}^n\right)\right)\times\left(-\mathbf{C}_n^b\mathbf{g}^n\right) = \mathbf{C}_n^b\left(\left(\mathbf{g}^n\times\boldsymbol{\Omega}^n\right)\times\mathbf{g}^n\right)$$

We could use the above identities to show that

$$\left((\hat{\mathbf{a}}\times\hat{\boldsymbol{\omega}})\times\hat{\mathbf{a}} \quad -\hat{\mathbf{a}}\times\hat{\boldsymbol{\omega}} \quad -\hat{\mathbf{a}}\right) = \left(\mathbf{C}_n^b\left(\left(\mathbf{g}^n\times\boldsymbol{\Omega}^n\right)\times\mathbf{g}^n\right) \quad \mathbf{C}_n^b\left(\mathbf{g}^n\times\boldsymbol{\Omega}^n\right) \quad \mathbf{C}_n^b\mathbf{g}^n\right)$$
$$= \mathbf{C}_n^b\left(\left(\mathbf{g}^n\times\boldsymbol{\Omega}^n\right)\times\mathbf{g}^n \quad \mathbf{g}^n\times\boldsymbol{\Omega}^n \quad \mathbf{g}^n\right)$$

which implies that

$$\mathbf{C}_n^b = \left((\hat{\mathbf{a}} \times \hat{\boldsymbol{\omega}}) \times \hat{\mathbf{a}} \quad -\hat{\mathbf{a}} \times \hat{\boldsymbol{\omega}} \quad \text{-}\hat{\mathbf{a}}\right)\left(\left(\mathbf{g}^n \times \boldsymbol{\Omega}^n\right) \times \mathbf{g}^n \quad \mathbf{g}^n \times \boldsymbol{\Omega}^n \quad \mathbf{g}^n\right)^{-1}$$

Finally, to find the new estimate for $\mathbf{C}_n^b$ we use Eq. (9.6) and substitute from the above to get

$$\begin{aligned}\hat{\mathbf{C}}_n^b &= \left((\hat{\mathbf{a}} \times \hat{\boldsymbol{\omega}}) \times \hat{\mathbf{a}} \quad -\hat{\mathbf{a}} \times \hat{\boldsymbol{\omega}} \quad \text{-}\hat{\mathbf{a}}\right)\left((\mathbf{g}^n \times \boldsymbol{\Omega}^n) \times \mathbf{g}^n \quad \mathbf{g}^n \times \boldsymbol{\Omega}^n \quad \mathbf{g}^n\right)^{-1} \\ &= \mathbf{C}_n^b\end{aligned} \tag{9.39}$$

This shows that the original estimate for $\mathbf{C}_n^b$ does not change with knowledge of sensor biases.

In this chapter we illustrated the alignment process using highly accurate sensors. In the following chapter we introduce a different process for alignment when less accurate sensors are used.

References

1. C. Jekeli, Inertial Navigation Systems with Geodetic Applications, Walter de Gruyter, Berlin, 2001.
2. J. A. Farrell & M. Barth, The Global Positioning System & Inertial Navigation, McGraw-Hill, New York, 1999.
3. R. Dorobantu, C. Gerlach, Investigation of a Navigation-Grade RLG SIMU type iNav-RQH,' IAPG/ FESG No. 16, München 2004. http://www.imar-navigation.de/download/inertial_navigation_introduction.pdf
4. K.R. Britting, Inertial Navigation Systems Analysis, Wiley-Interscience, 1971.
5. Yeon Fuh Jiang, Chung Shan, 'Error Analysis of Analytic Coarse Alignment Methods,' IEEE Transaction on Aerospace and Electronic Systems, Vol. 34, Jan. 1998, p334.
6. G. Strang, Linear Algebra and Applications, Saunders HBJ, Fort Worth, 1988.

Chapter 10

Attitude and Heading Reference System

10.1 Introduction

Attitude and heading reference system (AHRS) is the avionic system that enables the aircraft pilot to establish an artificial horizontal and bearing. Its function is to measure the attitude – pitch and roll – and trajectory heading for an aircraft. The legacy AHRS was built with vertical and directional gyros technology. More recently, modern AHRS uses micro electro-mechanical sensors (MEMS) gyros. Although with this technology gyro drifts could be so high that they are rendered unsuitable for navigation and alignment. And for this reason, an AHRS is usually aided with external navigational sensors like GPS, air data systems or magnetometers, even though there are some standalone AHRSs.

Since the attitude and heading data are inherent in $\mathbf{C}_b^n$ or $\mathbf{Q}_b^n$, they must be properly initialized. On the ground, craft attitude is typically determined from accelerometer sensor data. Heading can be initialized from magnetic sensors or, when flying, from GPS. In the meantime, for reliable operation, gyro drifts must be accounted for and should be continually estimated.

Herein we address the initialization procedure and provide the means for the detection and estimation of the low-grade gyro drifts.

10.2 Attitude Initialization

As before, we assume that each of the gyro and accelerometer triads comprise mutually orthogonal sensors that are perfectly aligned with the craft's principal body axes.

Here, $\mathbf{C}_n^b$ is considered the product of the Euler angle rotations

$$\mathbf{C}_n^b = [\phi]_x[\theta]_y[\psi]_z \tag{10.1}$$

In the following, we describe the procedure for initializing the Euler angles (ϕ, θ, ψ). In contrast to the alignment procedure described in the previous chapter, here we don't compute $\mathbf{C}_n^b$ as a whole but rather in two steps: estimating the attitude angles (ϕ, θ), then estimating the heading angle ψ. The first step utilizes the accelerometer measurements for performing the computations as follows. Since the gravity vector, $\mathbf{a}^n$, in NED frame is given by

$$\mathbf{a}^n = \begin{bmatrix} 0 & 0 & -g \end{bmatrix}' \tag{10.2}$$

then the accelerometer measurement vector, $\mathbf{a}^b$, is given by

$$\begin{bmatrix} a_x \\ a_y \\ a_z \end{bmatrix} = \mathbf{a}^b = \mathbf{C}_n^b \mathbf{a}^n = \mathrm{g} \begin{bmatrix} s\theta \\ -c\theta\ s\phi \\ -c\theta\ c\phi \end{bmatrix} \tag{10.3}$$

Solving for the attitude angles in the above equation yields

$$\begin{bmatrix} \mathrm{s}\phi \\ \mathrm{c}\phi \end{bmatrix} = \frac{1}{\sqrt{a_y^2 + a_z^2}} \begin{bmatrix} -a_y \\ -a_z \end{bmatrix} \tag{10.4}$$

$$\begin{bmatrix} \mathrm{s}\theta \\ \mathrm{c}\theta \end{bmatrix} = \frac{1}{g} \begin{bmatrix} a_x \\ -(a_y s\phi + a_z c\phi) \end{bmatrix} \tag{10.5}$$

It should be noted that determining the sines and cosines of the Euler angles (ϕ, θ) is sufficient to compute the transformation matrix in Eq. (10.1); that is there is no need to evaluate the angles explicitly. If the quaternions are used then their initial state is determined by

$$\mathbf{q}_n^b = \mathbf{q}_z\left[\frac{\psi}{2}\right]\mathbf{q}_y\left[\frac{\theta}{2}\right]\mathbf{q}_x\left[\frac{\phi}{2}\right] = \begin{bmatrix} \mathrm{c}\dfrac{\psi}{2} \\ 0 \\ 0 \\ \mathrm{s}\dfrac{\psi}{2} \end{bmatrix}\begin{bmatrix} \mathrm{c}\dfrac{\theta}{2} \\ 0 \\ \mathrm{s}\dfrac{\theta}{2} \\ 0 \end{bmatrix}\begin{bmatrix} \mathrm{c}\dfrac{\phi}{2} \\ \mathrm{s}\dfrac{\phi}{2} \\ 0 \\ 0 \end{bmatrix} \tag{10.6}$$

Similarly, there is no need for explicit angle computations. One way to compute the sines and cosines of the half angles is through the trigonometric identities:

$$\begin{aligned} \mathrm{c}\frac{\theta}{2} &= \sqrt{\frac{1+\mathrm{c}\theta}{2}} \\ \mathrm{s}\frac{\theta}{2} &= \frac{\mathrm{s}\theta}{2\mathrm{c}\dfrac{\theta}{2}} \end{aligned} \tag{10.7}$$

What remains is to determine the heading angle, ψ, to completely compute Eqs. (10.1) and (10.6). As mentioned earlier, heading is determined from an external source such as magnetic sensors or GPS when the craft is flying. We describe in the following the process for computing the heading, using a 3-axis magnetic sensor. This process can be used on ground or during flight.

10.3 Heading Initialization

The Earth's magnetic field [1-2], is characterized by its magnetic intensities, (X, Y, Z), along the NED frame, as shown in Fig. 10.1. The horizontal intensity, H, is the vector sum of the X and Y components. Along its direction lies the magnetic north in which a frictionless suspended magnetic needle would align itself into. The deviation of the magnetic north from the geographic north is called the magnetic variation, or declination angle, D. The angle between the the horizontal intensity, H, and the total intensity, F, is the inclination angle, I. The value of the declination angle can obtained by the World Magnetic Model (WMM) 2000 as a function of the latitude and longitude [2]. It should be noted that the declination angle not only changes with location but also with time and should be updated periodically to allow for reliable navigation.

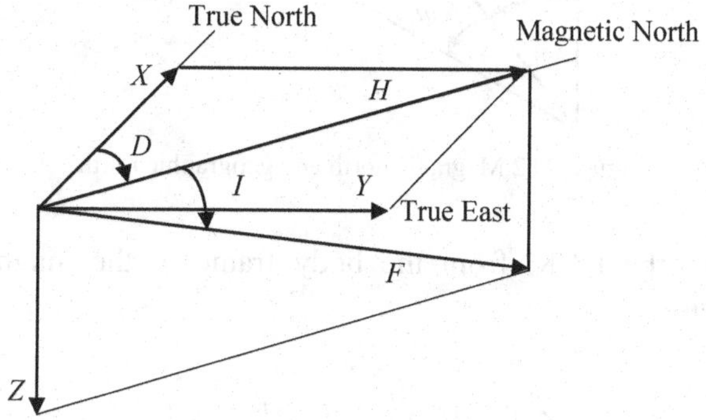

Figure 10.1 The Earth's Magnetic Elements

In the following we see how the magnetic intensity vector can be used to determine the craft's heading. A magnetic sensor is 3-axis magnetometer that measures the components of the magnetic field on its own coordinate frame. If the magnetometer is mounted so that its axes are aligned with the NED frame, its measurement vector would comprise the (X, Y, Z) components of the Earth magnetic field vector. However, in aircraft the magnetometer is, ideally, mounted so that its axes are aligned to the craft principal axes, and as such its measurements are in the craft body. We observe that if the craft is level and its x-axis is pointing towards the magnetic north then the magnetometer would sense the horizontal, H, and the vertical, Z, intensities of the magnetic field along its x and z axes, respectively, and the y component would be null.

In general, the craft attitude is deviated from the NED by the roll, pitch and yaw angles, (ϕ, θ, ψ). For convenience we introduce a new frame called the magnetic frame, denoted by m. This frame is obtained by rotating the NED frame about the vertical axis by the magnetic variation angle, see Fig.10.2.

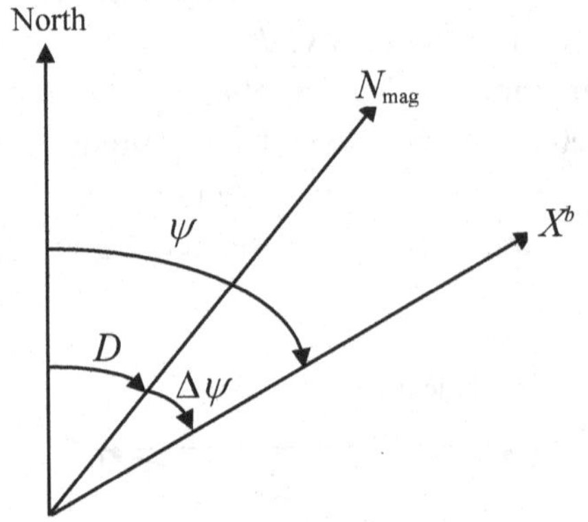

Figure 10.2 Magnetic north and geographic north

Therefore the DCM from the body frame to the magnetic frame, $\mathbf{C}_b^m$, is given by

$$\mathbf{C}_b^m = [-\Delta\psi]_z [-\theta]_y [-\phi]_x \tag{10.8}$$

where

$$\Delta\psi = \psi - D \tag{10.9}$$

Now, the sensor measured vector, $\mathbf{m}^b$, is related to the Earth's magnetic field vector, $\mathbf{m}^m$, by

$$\begin{bmatrix} H \\ 0 \\ Z \end{bmatrix} = \mathbf{m}^m = \mathbf{C}_b^m \mathbf{m}^b \tag{10.10}$$

Substituting from Eq. (10.8) into (10.10) yields

$$[\Delta\psi]_z \, \mathbf{m}^m = [-\theta]_y \, [-\phi]_x \, \mathbf{m}^b \tag{10.11}$$

Expanding the rotation matrices in the above equation yields

$$\begin{bmatrix} \mathrm{c}\Delta\psi & \mathrm{s}\Delta\psi & 0 \\ -\mathrm{s}\Delta\psi & \mathrm{c}\Delta\psi & 0 \\ 0 & 0 & 1 \end{bmatrix} \begin{bmatrix} H \\ 0 \\ Z \end{bmatrix} = \begin{bmatrix} \mathrm{c}\theta & 0 & \mathrm{s}\theta \\ 0 & 1 & 0 \\ -\mathrm{s}\theta & 0 & \mathrm{c}\theta \end{bmatrix} \begin{bmatrix} 1 & 0 & 0 \\ 0 & \mathrm{c}\phi & -\mathrm{s}\phi \\ 0 & \mathrm{s}\phi & \mathrm{c}\phi \end{bmatrix} \begin{bmatrix} m_x^b \\ m_y^b \\ m_z^b \end{bmatrix} \tag{10.12}$$

and by multiplying the terms we get

$$\begin{bmatrix} \mathrm{c}\,\Delta\psi\ H \\ -\mathrm{s}\,\Delta\psi\ H \\ Z \end{bmatrix} = \begin{bmatrix} \mathrm{c}\theta\ m_x^b + \mathrm{s}\theta\ \mathrm{s}\phi\ m_y^b + \mathrm{s}\theta\ \mathrm{c}\phi\ m_z^b \\ \mathrm{c}\phi\ m_y^b - \mathrm{s}\phi\ m_z^b \\ -\mathrm{s}\theta\ m_x^b + \mathrm{c}\theta\ \mathrm{s}\phi\ m_y^b + \mathrm{c}\theta\ \mathrm{c}\phi\ m_z^b \end{bmatrix} \tag{10.13}$$

From the first two components of the above equation we can write

$$\tan \Delta\psi = \frac{-\mathrm{c}\phi\ m_y^b + \mathrm{s}\phi\ m_z^b}{\mathrm{c}\theta\ m_x^b + \mathrm{s}\theta\ \mathrm{s}\phi\ m_y^b + \mathrm{s}\theta\ \mathrm{c}\phi\ m_z^b} \tag{10.14}$$

Substituting for the trigonometric terms from Eqs. (10.4)-(10.5) in the above equation and using (10.9), the heading can be determined completely. Alternatively, raw accelerometer data can be used in lieu of the attitude data, as follows. For simplification, we normalize the accelerometer measurements by the Earth gravity, i.e.

$$\begin{bmatrix} \bar{a}_x \\ \bar{a}_y \\ \bar{a}_z \end{bmatrix} = \frac{1}{g} \begin{bmatrix} a_x \\ a_y \\ a_z \end{bmatrix} \tag{10.15}$$

In terms of normalized accelerations Eqs. (10.4)-(10.5) become

$$\begin{bmatrix} s\phi \\ c\phi \end{bmatrix} = \frac{1}{\sqrt{\bar{a}_y^2 + \bar{a}_z^2}} \begin{bmatrix} -\bar{a}_y \\ -\bar{a}_z \end{bmatrix} \tag{10.16}$$

$$\begin{bmatrix} s\theta \\ c\theta \end{bmatrix} = \begin{bmatrix} \bar{a}_x \\ \sqrt{\bar{a}_y^2 + \bar{a}_z^2} \end{bmatrix} \tag{10.17}$$

Substituting from Eqs. (10.16) and (10.17) into Eq. (10.14)

$$\tan \Delta\psi = \frac{\left(\bar{a}_z m_y^b - \bar{a}_y m_z^b\right)}{\left(\bar{a}_y^2 + \bar{a}_z^2\right) m_x^b - \bar{a}_x \bar{a}_y m_y^b - \bar{a}_x \bar{a}_z m_z^b} \tag{10.18}$$

Upon collecting the terms in the denominator, the above equation becomes

$$\tan \Delta\psi = \frac{\left(\bar{a}_z m_y^b - \bar{a}_y m_z^b\right)}{\bar{a}_y\left(\bar{a}_y m_x^b - \bar{a}_x m_y^b\right) - \bar{a}_z\left(\bar{a}_x m_z^b - \bar{a}_z m_x^b\right)} \tag{10.19}$$

Notice that the cross product of the magnetometer and the accelerometer vectors is

$$\begin{bmatrix} p_x \\ p_y \\ p_z \end{bmatrix} = \begin{bmatrix} m_x^b \\ m_y^b \\ m_z^b \end{bmatrix} \times \begin{bmatrix} \bar{a}_x \\ \bar{a}_y \\ \bar{a}_z \end{bmatrix} = \begin{bmatrix} \bar{a}_z m_y^b - \bar{a}_y m_z^b \\ \bar{a}_x m_z^b - \bar{a}_z m_x^b \\ \bar{a}_y m_x^b - \bar{a}_x m_y^b \end{bmatrix} \tag{10.20}$$

Hence, the heading equation can be simplified to

$$\tan \Delta\psi = \frac{p_x}{\bar{a}_y p_z - \bar{a}_z p_y} \tag{10.21}$$

Computing the arctangent of Eq.(10.21) and using the WMM to compute the magnetic declination angle, then from Eq. (10.9), we arrive at the true heading angle as

$$\psi = \Delta\psi + D \tag{10.22}$$

10.4 Gyro Drift Compensation

Even though gyros are calibrated and compensated for biases, they exhibit additional errors due to “drift”. Gyro drift amounts to departure of the gyro bias from its calibrated value. Naturally, if these drifts are not compensated for, they will cause major errors in estimated attitude and heading angles.

Herein we introduce two algorithms for gyro bias compensation; the first uses the traditional g-slaving, and the other is based on the body Nav-frame DCM error. Regardless of which method is used for corrections, it is worth noting that only during steady and level flight do we detect and estimate gyro drifts. During accelerations caused by linear or rotational maneuvers, no drift detection is attempted as this can cause severe errors due to mixing of actual errors with real accelerations. Thus, gyro bias estimates detected during steady flight will be used to correct for gyro drifts to enable for more accurate attitude and heading estimation. In the following analysis we adopt the NED notation for both the navigation and body frames.

10.5 G Slaving

This approach is based on the observation: an INS computer on board of a non-accelerating (non-maneuvering) craft that uses drift free gyro data will not exhibit horizontal accelerations. Unaccounted gyro biases (drifts) will then cause $\mathbf{C}_b^n$ to show false rotations. Consequently, the acceleration computed in the Nav-frame will detect components of the gravity in the horizontal plane. The idea is to use these "false" components to correct for errors. It will be noticed that the g-slaving can estimate gyro errors in the roll and pitch channels only. We now analyze the effect of each respective gyro bias.

10.5.1 X-Gyro Bias

A positive bias in x-gyro causes the body frame to appear as if rotated about the x^b-axis by an angle, say, $\delta\phi$. This angle is dependent on the amount of drift and the duration of observation. In turn, this false body rotation will cause the computed accelerometer triad to also appear rotated with the body. However actual accelerometer measurements will not be affected by the false body rotation. Specifically, the y-axis accelerometer will appear to have been tilted by an angle $\delta\phi$ in the vertical plane but the y-accelerometer measurement, provided that it is originally horizontal, will be null. When accelerometer measurements are transformed from the body to the Nav-frame an acceleration component, a_y^n=gs($\delta\phi$), proportional to the tilt angle, $\delta\phi$, will be detected along the y^n-axis (as shown in Fig. 10.3).

We could have arrived at this result mathematically by noticing that

$$\mathbf{a}^n = \mathbf{C}_b^n \mathbf{a}^b = \begin{bmatrix} 1 & 0 & 0 \\ 0 & \mathrm{c}\delta\phi & -\mathrm{s}\delta\phi \\ 0 & \mathrm{s}\delta\phi & \mathrm{c}\delta\phi \end{bmatrix} \begin{bmatrix} 0 \\ 0 \\ -g \end{bmatrix}$$

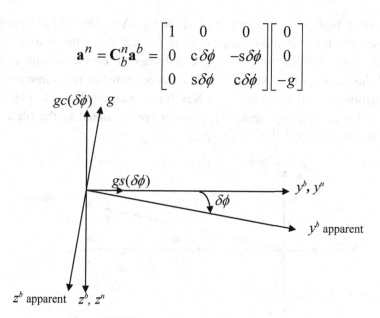

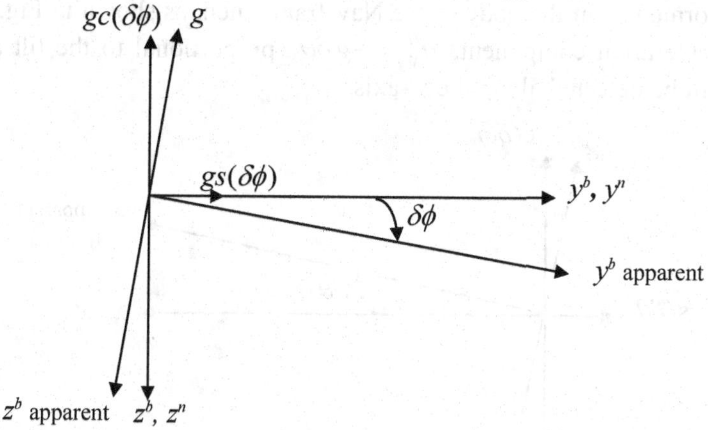

Figure 10.3 Effects of *x*-gyro drift on acceleration

which implies that

$$\mathbf{a}^n = g \begin{bmatrix} 0 \\ \mathrm{s}\delta\phi \\ -\mathrm{c}\,\delta\phi \end{bmatrix} \tag{10.23}$$

Thus a positive tilt about the *x*-axis results in a positive gravity component along the *y*-axis. Provided that $\delta\phi$ is small, Eq. (10.23) gives

$$\delta\phi = \frac{a_y^n}{g} \tag{10.24}$$

To correct for this false rotation, the body frame must be rotated opposite to the angle indicated by a_y^n.

10.5.2 Y-Gyro Bias

We treat bias error in the *y*-gyro similarly. A positive bias in y-gyro causes the body frame to appear as if rotated about the y^b-axis by an angle, say, $\delta\theta$. Likewise, this angle is dependent on the amount of drift and the duration of observation. When accelerometer measurements are transformed from the body to the Nav-frame then, as shown in Fig. 10.4, an acceleration component, $a_x^n = -g\delta\theta$, proportional to the tilt angle, $\delta\theta$, will be detected along the x^n-axis.

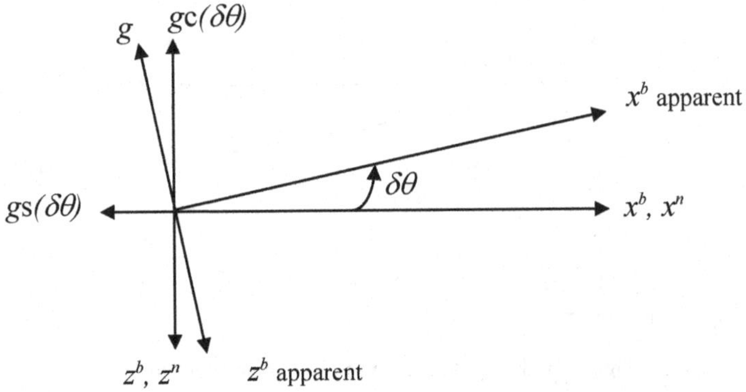

Figure 10.4 Effects of *y*-gyro drift on acceleration

This result can be derived mathematically, by noting that

$$\mathbf{a}^n = \mathbf{C}_b^n \mathbf{a}^b = \begin{bmatrix} c\delta\theta & 0 & s\delta\theta \\ 0 & 1 & 0 \\ -s\delta\theta & 0 & c\delta\theta \end{bmatrix} \begin{bmatrix} 0 \\ 0 \\ -g \end{bmatrix}$$

which implies that

$$\mathbf{a}^n = g\begin{bmatrix} -\mathrm{s}\ \delta\theta \\ 0 \\ -\mathrm{c}\ \delta\theta \end{bmatrix} \tag{10.25}$$

Thus a positive tilt about the *y*-axis results in negative gravity component along the *y*-axis. Provided that $\delta\theta$ is small, Eq.(10.25) gives

$$\delta\theta = -\frac{a_x^n}{g} \tag{10.26}$$

To correct for this false rotation, the body frame must be rotated opposite to the angle indicated by a_x^n.

10.5.3 Z-Gyro Bias

A positive bias in *z*-gyro will cause the body frame to rotate about the *z*-axis. However accelerometer measurements cannot detect vertical axis gyro tilt. An external source such as a magnetometer or GPS must be used to detect such tilts. If $\overline{\psi}$ is the heading observed by solving the navigation equation and ψ is the reference (magnetometer or GPS) observation, then the error $\delta\psi = \overline{\psi} - \psi$ represents the bias in the *z*-gyro. The body frame must be tilted by $-\delta\psi$ about the *z*-axis.

10.6 Alternative Approach for Gyro Drift Compensation

This approach is based on the observation that measured accelerations do not actually contribute to the process of gyro drift. As such, one might consider the idea of computing the tilts from $\mathbf{C}_b^n$ and use them to update the errors in gyro bias.

When a craft flies straight and level then ideally its $\mathbf{C}_b^n$ is given by

$$\left(\mathbf{C}_b^n\right)_{ideal} = \begin{bmatrix} c\psi & -s\psi & 0 \\ s\psi & c\psi & 0 \\ 0 & 0 & 1 \end{bmatrix} \tag{10.27}$$

where the heading angle is given by magnetometer or GPS. When the tilts are small, then small angle approximation gives

$$\delta\mathbf{C}_b^n = \left(\mathbf{C}_n^b\right)_{ideal}\left(\mathbf{C}_b^n\right)_{computed} = \begin{bmatrix} 1 & -\delta\psi & \delta\theta \\ \delta\psi & 1 & -\delta\phi \\ -\delta\theta & \delta\phi & 1 \end{bmatrix} \tag{10.28}$$

from which tilts can be computed by

$$\begin{bmatrix} \delta\phi \\ \delta\theta \\ \delta\psi \end{bmatrix} = \frac{1}{2}\begin{bmatrix} \mathbf{C}_{b(3,2)}^n - \mathbf{C}_{b(2,3)}^n \\ \mathbf{C}_{b(1,3)}^n - \mathbf{C}_{b(3,1)}^n \\ \mathbf{C}_{b(2,1)}^n - \mathbf{C}_{b(1,2)}^n \end{bmatrix} \tag{10.29}$$

Finally, gyro drifts are estimated by averaging the ramps in error angles (in Eqs. (10.24), (10.26) in the g-slaving approach or in Eq. (10.29)) over the time intervals in which they are observed.

10.7 Maneuver Detector

To implement either of the above approaches, we need to identify two mutually distinct modes of flight. The first is the non-accelerated mode in which a craft flies steady and level at constant speed and during which gyro drifts can be estimated and corrected for. The other is the accelerated mode in which the craft accelerates linearly or rotationally about any of its axes. In this mode no drifts are estimated, and only prior estimates are used to correct for gyro drifts.

Thus it is necessary to establish a maneuver detector to distinguish between these two modes. Possibly there could be a gray zone in which corrections are reduced. The maneuver detector could be governed by the following equations

$$\begin{aligned} \left\|\boldsymbol{\omega}_{ib}^{b}\right\| &\le Th_r \\ \left\|\mathbf{a}^{b}\right\| &\le Th_a \end{aligned} \qquad (10.30)$$

For a rate gyro, the threshold should be selected to fall between two levels, as depicted in Fig. 10.5:

a. The maximum rate gyro noise level or the maximum craft disturbances.

b. The minimum possible sustained maneuver above disturbances

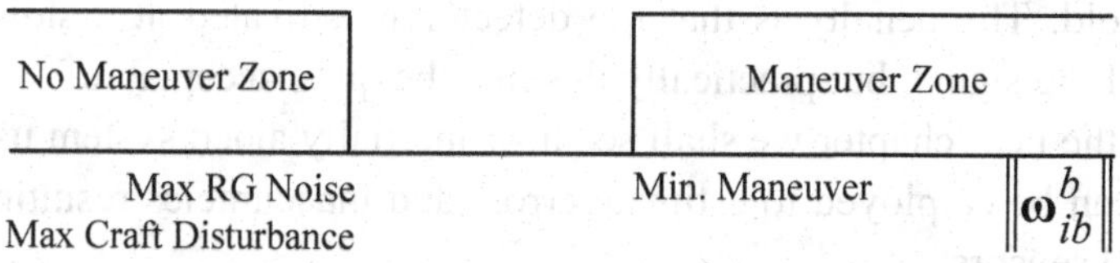

Figure 10.5 Selection of Maneuver Detection Thresholds

10.7.1 Rate Gyro Threshold Selection

First we note that the standard deviation of a sampled band-limited analog signal is defined by two parameters:

a. Sensor noise power spectral density, Φ (deg/sec)^2/Hz, or voltage N deg/sec/√Hz
b. Sampling rate, s, samples/sec.

Thus, if a rate gyro noise voltage is N deg /sec/√Hz and its data are sampled at s samples/sec, then a one-sigma standard deviation for the sampled data is

$$1\sigma = N\sqrt{s}\ \text{deg/sec}$$

Assuming for example that N=180 deg/hour/√Hz=.05 deg/sec/√Hz, and that the rate gyro signal is sampled at a rate, s=400 Hz, then 1σ=.05√400 deg/sec. The root sum square, σ_{rss} (considering 3 orthogonal gyros) is given by $1\sigma_{rss}$ =√3.1 deg/sec =1.73 deg/sec.

The selection of the threshold depends on many factors such as the craft's operational environment, the sensors quality, the sampling rate and the acceptable tolerances of errors. The following provides an example of how the threshold may be selected. Often, a reasonable choice of the threshold is Th_r=2σ_{rss}=2(1.73) deg/sec =3.46 deg/sec.

This threshold may be much higher than the minimum sustainable maneuver. In this case, the rate gyro signal could be averaged, say over a period of one second. This implies that 400 data samples will used to average the threshold. Hence, the new threshold for the averaged signal becomes Th_r=3.46/√400 deg/sec =.173 deg/sec.

This shows that signal averaging is a useful tool for manipulating the threshold. The penalty is that the detector is activated at a slower rate than of the signal, but practically this may be quite acceptable.

In the next chapter we shall see how inertially aided system using the GPS can be employed to estimate errors and inaccuracies resulting from inertial sensors.

References

1. Further Understanding of Geomagnetism, http://www.ngdc.noaa.gov/seg/geomag/geomaginfo.shtml
2. World Magnetic Model - Fortran Download

Chapter 11

GPS Aided Inertial System

11.1 Introduction

Errors in the navigation solution of position and velocity can build up due to unaccounted for sensor noise and error sources. Chief amongst them are gyro drifts and accelerometer biases. The g-slaving method was used with AHRS to estimate these errors, however this method can not be used when the craft maneuvers or can it reestablish any reference point after the craft ends the maneuver. An extra navigation aid is needed to establish the reference points and estimate the inertial errors.

The Global Positioning System (GPS) is one helpful aid for estimating these error sources. It provides, with a reasonable degree of accuracy, the position, the velocity and the heading of the craft. Thus it can establish a reference navigation point after the period in which the inertial navigation system (INS) has been acting alone. To employ such aiding tool, a GPS receiver is installed and a Kalman filter (KF) is designed to estimate the navigational error parameters. A processor on board receives and compares the position and velocity data from both the INS and the GPS. The residual – the difference between the INS and GPS navigation data – is weighted by the KF gain to correct the error in the INS estimates. Among the many approaches for implementing the KF, we herein select for our discussion the loosely coupled algorithm. This approach is selected for its simplicity in that it does not need to access GPS receiver internal raw data and the processing of the INS and GPS data are completely decoupled.

Three elements are needed to implement the aiding tool:

i. The INS interface with the Kalman filter: this comprises the error state model and the position and velocity navigation data as computed by the INS. At minimum, the state error should include the position, velocity, attitude, gyro drift, and accelerometer bias errors. The state is the vector that comprises the elements that we desire to estimate and its dynamic equation that describes how these elements propagate in time.

ii. The GPS interface with the Kalman Filter: It is essentially the craft position and velocity navigation data and some frame transformation if the data are computed in a coordinate system different than that of the INS.

iii. The Kalman filter interface with both the INS and GPS. This is a complete implementation of the KF equations for processing the navigation data and estimating the INS errors. The measurement, or observation, equation relates the GPS measurements to the state vector. The observation vector here will be the difference between the GPS and the INS craft's three-dimensional position. Because the state equation in our application is non linear, we implement a special form of KF, called extended KF, in which it estimates the error in the state rather than the state itself. The error in the vector velocity could be augmented if the GPS receiver provides velocity measurements.

In the following we address the above components. We start by deriving the INS state error equations. These equations are derived in two frames, the navigation and Earth frames.

11.2 Navigation Frame Error Equation

To simplify derivations, we discuss first the contribution of each navigational variable, derive the dynamic equation of each these component, and then integrate all the equations to obtain the state vector dynamic equation. The following vectors denote the error in craft position and velocity, respectively

$$\delta \mathbf{p} = \begin{bmatrix} \delta\phi \\ \delta\lambda \\ \delta h \end{bmatrix} \tag{11.1}$$

$$\delta \mathbf{v}^n = \begin{bmatrix} \delta v_n \\ \delta v_e \\ \delta v_z \end{bmatrix} \tag{11.2}$$

11.2.1 Craft Rate Error $\delta\boldsymbol{\omega}_{en}^n$

The craft rate, given by Eq. (4.11), is

$$\boldsymbol{\omega}_{en}^n = \begin{bmatrix} \dfrac{v_e}{R_p + h} \\ -\dfrac{v_n}{R_m + h} \\ -\dfrac{v_e \tan\phi}{R_p + h} \end{bmatrix} \tag{11.3}$$

implies that

$$\delta\boldsymbol{\omega}_{en}^n = \begin{bmatrix} -\dfrac{v_e \delta h}{(R_p + h)^2} \\ \dfrac{v_n \delta h}{(R_m + h)^2} \\ \dfrac{v_e \tan(\phi)\delta h}{(R_p + h)^2} - \dfrac{v_e \sec^2(\phi)\delta\phi}{R_p + h} \end{bmatrix} + \begin{bmatrix} 0 & \dfrac{1}{R_p + h} & 0 \\ -\dfrac{1}{R_m + h} & 0 & 0 \\ 0 & -\dfrac{\tan\phi}{R_p + h} & 0 \end{bmatrix} \begin{bmatrix} \delta v_n \\ \delta v_e \\ \delta v_z \end{bmatrix} \tag{11.4}$$

By ignoring the altitude and assuming that the Earth's principal and meridional radii are equal to their geometric mean, R, the reciprocal Earth radius of curvature terms will all become equal and as such the above equation can be simplified into

$$\delta\boldsymbol{\omega}_{en}^{n} = \mathbf{C}_{\omega np}\delta\mathbf{p} + \mathbf{C}_{\omega nv}\delta\mathbf{v}^{n} \tag{11.5}$$

where

$$\mathbf{C}_{\omega np} = \begin{bmatrix} 0 & 0 & -\dfrac{v_e}{(R_p+h)^2} \\ 0 & 0 & \dfrac{v_n}{(R_m+h)^2} \\ -\dot{\lambda}\sec\phi & 0 & \dfrac{v_e \tan\phi}{(R_p+h)^2} \end{bmatrix} \tag{11.6}$$

and

$$\mathbf{C}_{\omega nv} = \begin{bmatrix} 0 & \dfrac{1}{R_p+h} & 0 \\ -\dfrac{1}{R_m+h} & 0 & 0 \\ 0 & -\dfrac{\tan\phi}{R_p+h} & 0 \end{bmatrix} \tag{11.7}$$

11.2.2 Earth Rate Error $\delta\boldsymbol{\omega}_{ie}^{n}$

The Earth rate, in Eq. (4.7), is given by

$$\boldsymbol{\omega}_{ie}^{n} = \begin{bmatrix} \cos\phi \\ 0 \\ -\sin\phi \end{bmatrix} \Omega_e \tag{11.8}$$

Assume that the Earth rate, Ω_e, is constant implies

$$\delta\boldsymbol{\omega}_{ie}^{n} = \begin{bmatrix} -\Omega_e \sin\phi \\ 0 \\ -\Omega_e \cos\phi \end{bmatrix} \delta\phi \tag{11.9}$$

$$\delta\boldsymbol{\omega}_{ie}^{n} = \mathbf{C}_{\omega ep} \delta\mathbf{p} \tag{11.10}$$

where

$$\mathbf{C}_{\omega ep} = \begin{bmatrix} -\Omega_e \sin\phi & & \\ 0 & \mathbf{0} & \mathbf{0} \\ -\Omega_e \cos\phi & & \end{bmatrix} \tag{11.11}$$

11.2.3 Position Errors

Equation (4.8) gives

$$\dot{\phi} = \frac{v_n}{R_m + h}$$

from which we get

$$\begin{aligned}\delta\dot{\phi} &= -\frac{v_n}{(R_m+h)^2}\delta h+\frac{1}{R_m+h}\delta v_n \\ &= -\frac{\dot{\phi}}{R_m+h}\delta h+\frac{1}{R_m+h}\delta v_n\end{aligned} \tag{11.12}$$

Likewise Eq. (4.9) gives

$$\dot{\lambda} = \frac{v_e}{(R_p+h)\cos\phi}$$

from which we get

$$\begin{aligned}\delta\dot{\lambda} &= \frac{\delta v_e}{(R_p+h)\cos\phi}+\frac{v_e\tan(\phi)\delta\phi}{(R_p+h)\cos\phi}-\frac{v_e\delta h}{(R_p+h)^2\cos\phi} \\ &= \dot{\lambda}\tan(\phi)\delta\phi-\frac{\dot{\lambda}\delta h}{R_p+h}+\frac{\dot{\lambda}}{v_e}\delta v_e\end{aligned} \tag{11.13}$$

The above two equations when combined with the vertical channel yield

$$\begin{bmatrix}\delta\dot{\phi}\\ \delta\dot{\lambda}\\ \delta\dot{h}\end{bmatrix} = \begin{bmatrix}0 & 0 & -\dfrac{\dot{\phi}}{R_m+h}\\ \dot{\lambda}\tan\phi & 0 & -\dfrac{\dot{\lambda}}{R_p+h}\\ 0 & 0 & 0\end{bmatrix}\begin{bmatrix}\delta\phi\\ \delta\lambda\\ \delta h\end{bmatrix}+\begin{bmatrix}\dfrac{1}{R_m+h} & 0 & 0\\ 0 & \dfrac{\dot{\lambda}}{v_e} & 0\\ 0 & 0 & -1\end{bmatrix}\begin{bmatrix}\delta v_n\\ \delta v_e\\ \delta v_z\end{bmatrix} \tag{11.14}$$

If

$$\mathbf{C}_{pp} = \begin{bmatrix} 0 & 0 & -\dfrac{\dot{\phi}}{R_m + h} \\ \dot{\lambda} \tan \phi & 0 & -\dfrac{\dot{\lambda}}{R_p + h} \\ 0 & 0 & 0 \end{bmatrix} \tag{11.15}$$

and

$$\mathbf{C}_{pv} = \begin{bmatrix} \dfrac{1}{R_m + h} & 0 & 0 \\ 0 & \dfrac{\dot{\lambda}}{v_e} & 0 \\ 0 & 0 & -1 \end{bmatrix} \tag{11.16}$$

then in a compact matrix form the craft position error vector is given by

$$\delta \dot{\mathbf{p}} = \mathbf{C}_{pp} \delta \mathbf{p} + \mathbf{C}_{pv} \delta \mathbf{v}^n \tag{11.17}$$

11.2.4 Attitude Error

We face a little challenge because the craft's attitude values are implicit in the body to navigation transformation matrix, $\mathbf{C}_b^n$, a 3×3 matrix and comprises nine elements. Since it is a DCM, it can be computed from three independent elements. The algebra of DCM errors, given in Appendix G, characterizes the error in a DCM in a three dimensional vector. This enables us to write the error vector, $\boldsymbol{\eta}$, in $\mathbf{C}_b^n$ as

$$\delta \mathbf{C}_b^n = -\tilde{\boldsymbol{\eta}} \mathbf{C}_b^n \tag{11.18}$$

and the dynamic equation of $\boldsymbol{\eta}$ will be given by

$$\dot{\boldsymbol{\eta}} = -\tilde{\boldsymbol{\omega}}_{bn}^n \boldsymbol{\eta} + \delta \boldsymbol{\omega}_{bn}^n \tag{11.19}$$

Equation (11.19) can be expressed in a form that relates explicitly to the errors desired to be estimated. Since

$$\boldsymbol{\omega}_{bn}^n = \boldsymbol{\omega}_{in}^n - \boldsymbol{\omega}_{ib}^n \tag{11.20}$$

then

$$\delta \boldsymbol{\omega}_{bn}^n = \delta \boldsymbol{\omega}_{in}^n - \delta \boldsymbol{\omega}_{ib}^n \tag{11.21}$$

Since

$$\begin{aligned} \delta \boldsymbol{\omega}_{ib}^n &= \delta \left(\mathbf{C}_b^n \, \boldsymbol{\omega}_{ib}^b \right) \\ &= \mathbf{C}_b^n \, \delta \boldsymbol{\omega}_{ib}^b + \left(\delta \mathbf{C}_b^n \right) \boldsymbol{\omega}_{ib}^b \end{aligned} \tag{11.22}$$

From Eq. (11.18) we see that

$$(\delta \mathbf{C}_b^n) \boldsymbol{\omega}_{ib}^b = -\tilde{\boldsymbol{\eta}} \mathbf{C}_b^n \boldsymbol{\omega}_{ib}^b = -\tilde{\boldsymbol{\eta}} \boldsymbol{\omega}_{ib}^n = \tilde{\boldsymbol{\omega}}_{ib}^n \boldsymbol{\eta}$$

Substituting for this term in Eq. (11.22) gives

$$\delta \boldsymbol{\omega}_{ib}^n = \mathbf{C}_b^n \, \delta \boldsymbol{\omega}_{ib}^b + \tilde{\boldsymbol{\omega}}_{ib}^n \boldsymbol{\eta} \tag{11.23}$$

Substituting Eq. (11.23) into Eq. (11.21) yields

$$\delta\boldsymbol{\omega}_{bn}^{n} = \delta\boldsymbol{\omega}_{in}^{n} - \mathbf{C}_{b}^{n}\,\delta\boldsymbol{\omega}_{ib}^{b} - \tilde{\boldsymbol{\omega}}_{ib}^{n}\boldsymbol{\eta} \tag{11.24}$$

Substituting Eqs. (11.20) and (11.24) into (11.19) yields

$$\dot{\boldsymbol{\eta}} = -\left(\tilde{\boldsymbol{\omega}}_{in}^{n} - \tilde{\boldsymbol{\omega}}_{ib}^{n}\right)\boldsymbol{\eta} + \delta\boldsymbol{\omega}_{in}^{n} - \mathbf{C}_{b}^{n}\,\delta\boldsymbol{\omega}_{ib}^{b} - \tilde{\boldsymbol{\omega}}_{ib}^{n}\boldsymbol{\eta}$$

which upon simplification becomes

$$\dot{\boldsymbol{\eta}} = -\tilde{\boldsymbol{\omega}}_{in}^{n}\boldsymbol{\eta} + \delta\boldsymbol{\omega}_{in}^{n} - \mathbf{C}_{b}^{n}\,\delta\boldsymbol{\omega}_{ib}^{b} \tag{11.25}$$

Notice, in the above equation, that $\delta\boldsymbol{\omega}_{ib}^{b}$ is the rate gyro drift errors. Since

$$\delta\boldsymbol{\omega}_{in}^{n} = \delta\boldsymbol{\omega}_{ie}^{n} + \delta\boldsymbol{\omega}_{en}^{n}$$

then from Eqs. (11.5) and (11.10)

$$\delta\boldsymbol{\omega}_{in}^{n} = (\mathbf{C}_{\omega np} + \mathbf{C}_{\omega ep})\delta\mathbf{p} + \mathbf{C}_{\omega nv}\delta\mathbf{v}^{n} \tag{11.26}$$

Substituting for the above in Eq. (11.25) yields

$$\dot{\boldsymbol{\eta}} = -\tilde{\boldsymbol{\omega}}_{in}^{n}\boldsymbol{\eta} + (\mathbf{C}_{\omega np} + \mathbf{C}_{\omega ep})\delta\mathbf{p} + \mathbf{C}_{\omega nv}\delta\mathbf{v}^{n} - \mathbf{C}_{b}^{n}\,\delta\boldsymbol{\omega}_{ib}^{b} \tag{11.27}$$

11.2.5 Gravity Error

We shall consider only errors in the downward normal gravity. Errors in gravity due to anomaly and deflection of the vertical will be treated as random. Also, we ignore the northward component because it is already of small value. Equation (4.21) shows that the normal gravity is given by

$$g = \gamma - \gamma \left(1 + f + m - 2f \ \sin^2 \phi\right) \frac{2h}{a}$$

from which

$$\delta g = g_\phi \, \delta\phi + g_h \, \delta h$$

where γ is given by Eqs. (4.17) and (4.18) and

$$g_\phi = \gamma' \left[1 - \frac{2}{a}\left(1 + f + m - 2f \ \sin^2 \phi\right) h\right] + \frac{4\gamma f \ h}{a} \sin(2\phi) \qquad (11.28)$$

$$\gamma' = \gamma_e \frac{\partial}{\partial\phi} \frac{1 + k \sin^2 \phi}{\sqrt{1 - e^2 \sin^2 \phi}} \qquad (11.29)$$

$$g_h = -\frac{2\gamma}{a}\left(1 + f + m - 2f \sin^2 \phi\right) \qquad (11.30)$$

Since

$$\mathbf{g}^n = \begin{bmatrix} 0 \\ 0 \\ g \end{bmatrix}$$

then

$$\delta \mathbf{g}^n = \begin{bmatrix} 0 \\ 0 \\ \delta g \end{bmatrix} = \begin{bmatrix} 0 \\ 0 \\ g_\phi \end{bmatrix} \delta\phi + \begin{bmatrix} 0 \\ 0 \\ g_h \end{bmatrix} \delta h \tag{11.31}$$

The gravity error, Eqs. (11.29)-(11.31), will be given by

$$\delta \mathbf{g}^n = \mathbf{G}_p \, \delta \mathbf{p} \tag{11.32}$$

where

$$\mathbf{G}_p = \begin{bmatrix} 0 & 0 & 0 \\ 0 & 0 & 0 \\ g_\phi & 0 & g_h \end{bmatrix} \tag{11.33}$$

11.2.6 Velocity Error

Let

$$\boldsymbol{\omega}^n = 2\boldsymbol{\omega}_{ie}^n + \boldsymbol{\omega}_{en}^n \tag{11.34}$$

then the acceleration as given by Eq. (5.17) becomes

$$\dot{\mathbf{v}}^n = \mathbf{C}_b^n \mathbf{a}^b - \boldsymbol{\omega}^n \times \mathbf{v}^n + \mathbf{g}^n$$

which implies that

$$\delta\dot{\mathbf{v}}^n = (\delta\mathbf{C}_b^n)\mathbf{a}^b + \mathbf{C}_b^n\delta\mathbf{a}^b - \delta\boldsymbol{\omega}^n \times \mathbf{v}^n - \boldsymbol{\omega}^n \times \delta\mathbf{v}^n + \delta\mathbf{g}^n$$

We notice from Eq. (11.18) that $(\delta\mathbf{C}_b^n)\mathbf{a}^b = -\tilde{\boldsymbol{\eta}}\mathbf{C}_b^n\mathbf{a}^b = -\tilde{\boldsymbol{\eta}}\mathbf{a}^n = \tilde{\mathbf{a}}^n\boldsymbol{\eta}$. Substituting for this term in the above equation, changing the cross products to skew symmetric matrices and rearranging terms yields

$$\delta\dot{\mathbf{v}}^n = \tilde{\mathbf{v}}^n\delta\boldsymbol{\omega}^n + \delta\mathbf{g}^n - \tilde{\boldsymbol{\omega}}^n\delta\mathbf{v}^n + \tilde{\mathbf{a}}^n\boldsymbol{\eta} + \mathbf{C}_b^n\delta\mathbf{a}^b \qquad (11.35)$$

Since

$$\delta\boldsymbol{\omega}^n = 2\delta\boldsymbol{\omega}_{ie}^n + \delta\boldsymbol{\omega}_{en}^n$$

then from Eqs. (11.5) and (11.10)

$$\delta\boldsymbol{\omega}^n = \left(\mathbf{C}_{\omega np} + 2\mathbf{C}_{\omega ep}\right)\delta\mathbf{p} + \mathbf{C}_{\omega nv}\delta\mathbf{v}^n \qquad (11.36)$$

Substituting for the above and from Eq. (11.32) in the velocity equation (11.35) yields

$$\begin{aligned}\delta\dot{\mathbf{v}}^n = \tilde{\mathbf{v}}^n&\left[(\mathbf{C}_{\omega np} + 2\mathbf{C}_{\omega ep})\delta\mathbf{p} + \mathbf{C}_{\omega nv}\delta\mathbf{v}^n\right]\\ &+\mathbf{G}_p\,\delta\mathbf{p} - \tilde{\boldsymbol{\omega}}^n\delta\mathbf{v}^n + \tilde{\mathbf{a}}^n\boldsymbol{\eta} + \mathbf{C}_b^n\delta\mathbf{a}^b\end{aligned}$$

Collecting terms

$$\begin{aligned}\delta\dot{\mathbf{v}}^n = &\left[\tilde{\mathbf{v}}^n\left(\mathbf{C}_{\omega np} + 2\mathbf{C}_{\omega ep}\right) + \mathbf{G}_p\right]\delta\mathbf{p}\\ &+\left[\tilde{\mathbf{v}}^n\mathbf{C}_{\omega nv} - \tilde{\boldsymbol{\omega}}^n\right]\delta\mathbf{v}^n + \tilde{\mathbf{a}}^n\boldsymbol{\eta} + \mathbf{C}_b^n\delta\mathbf{a}^b\end{aligned} \qquad (11.37)$$

11.2.7 Navigation Frame Error State Equation

From Eqs. (11.17), (11.27) and (11.37) we can describe the error state dynamics by the equation

$$\frac{d}{dt}\begin{bmatrix}\delta\mathbf{p}\\ \delta\mathbf{v}^n\\ \boldsymbol{\eta}\\ \delta\boldsymbol{\omega}_{ib}^b\\ \delta\mathbf{a}^b\end{bmatrix}=\begin{bmatrix}\mathbf{C}_{pp} & \mathbf{C}_{pv} & \mathbf{0} & \mathbf{0} & \mathbf{0}\\ \tilde{\mathbf{v}}^n\left(\mathbf{C}_{\omega np}+2\mathbf{C}_{\omega ep}\right)+\mathbf{G}_p & \left(\tilde{\mathbf{v}}^n\mathbf{C}_{\omega nv}-\tilde{\boldsymbol{\omega}}^n\right) & \tilde{\mathbf{a}}^n & \mathbf{0} & \mathbf{C}_b^n\\ \left(\mathbf{C}_{\omega np}+\mathbf{C}_{\omega ep}\right) & \mathbf{C}_{\omega nv} & -\tilde{\boldsymbol{\omega}}_{in}^n & -\mathbf{C}_b^n & \mathbf{0}\\ \mathbf{0} & \mathbf{0} & \mathbf{0} & \mathbf{0} & \mathbf{0}\\ \mathbf{0} & \mathbf{0} & \mathbf{0} & \mathbf{0} & \mathbf{0}\end{bmatrix}\begin{bmatrix}\delta\mathbf{p}\\ \delta\mathbf{v}^n\\ \boldsymbol{\eta}\\ \delta\boldsymbol{\omega}_{ib}^b\\ \delta\mathbf{a}^b\end{bmatrix}\tag{11.38}$$

We remark in the above equation that,

1. The entries in the state transition matrix, including the zeros, are 3×3 matrices.
2. Representing the error coefficients as blocks enables us to delete any term determined to be so small to warrant its inclusion.
3. The errors in the IMU errors are treated as biases. If they are not, then it will not be hard to modify the transition matrix to reflect any changes.

11.2.8 Error Block Diagram

The flow diagram that depicts the interdependence of the above error sources is depicted in Fig 11.1. It is, to great extent, a mirror image of the actual signal flow diagram depicted in Fig. 5.1. Both have the inertial sensor signals as the deriving source and three set of state equations for the position, velocity and attitude.

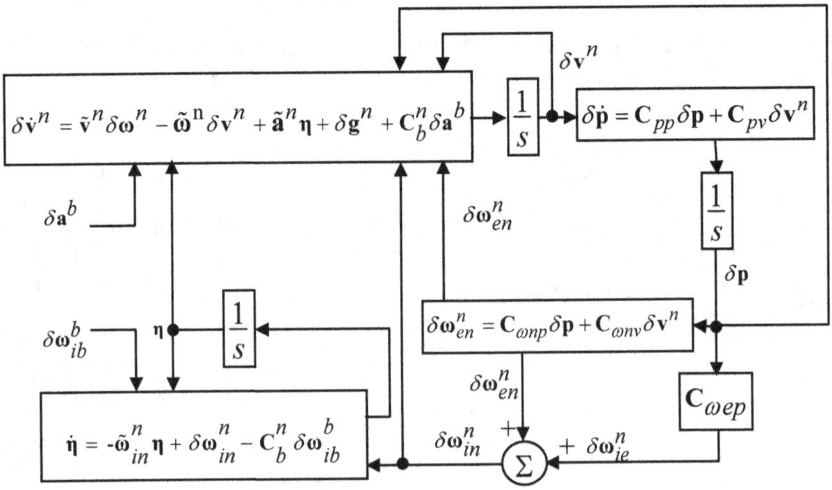

Figure 11.1 Computational Flow Diagram of the State Errors

11.3 Earth Frame Error Equations

In the navigation frame, both the position and velocity vectors were expressed in the Nav-frame. Alternatively, these two parameters could be expressed in an Earth Centered Earth fixed (ECEF), (a frame fixed to Earth and origin is at its center). Derivation of the error state is straightforward. The navigation state equation in the Earth frame is given in Chapter 5 by

$$\frac{d\mathbf{v}^e}{dt} = \mathbf{a}^e - 2\boldsymbol{\omega}_{ie}^e \times \mathbf{v}^e + \mathbf{g}^e$$

where

$$\boldsymbol{\omega}_{ie}^e = \begin{bmatrix} \Omega_e \\ 0 \\ 0 \end{bmatrix}$$

In this frame it is noticed that

$$\frac{d\mathbf{p}^e}{dt} = \mathbf{v}^e \tag{11.39}$$

11.3.1 Attitude Error

Similar to the attitude error in equations (11.18) and (11.19)

$$\delta\mathbf{C}_b^e = -\tilde{\varsigma}\mathbf{C}_b^e \tag{11.40}$$

and the dynamic equation of ς is given by

$$\dot{\varsigma} = -\tilde{\boldsymbol{\omega}}_{be}^e\varsigma + \delta\boldsymbol{\omega}_{be}^e \tag{11.41}$$

Notice that

$$\begin{aligned}\delta\boldsymbol{\omega}_{be}^e &= \delta\boldsymbol{\omega}_{ie}^e - \delta\boldsymbol{\omega}_{ib}^e \\ &= -\delta\left(\mathbf{C}_b^e\boldsymbol{\omega}_{ib}^b\right) \\ &= -\left(\delta\mathbf{C}_b^e\right)\boldsymbol{\omega}_{ib}^b - \mathbf{C}_b^e\delta\boldsymbol{\omega}_{ib}^b\end{aligned} \tag{11.42}$$

In the above it is assumed that the Earth rotation vector is constant. Substituting from Eq. (11.40)

$$\begin{aligned}\delta\boldsymbol{\omega}_{be}^e &= \tilde{\varsigma}\mathbf{C}_b^e\boldsymbol{\omega}_{ib}^b - \mathbf{C}_b^e\delta\boldsymbol{\omega}_{ib}^b \\ &= \tilde{\varsigma}\boldsymbol{\omega}_{ib}^e - \mathbf{C}_b^e\delta\boldsymbol{\omega}_{ib}^b \\ &= -\tilde{\boldsymbol{\omega}}_{ib}^e\varsigma - \mathbf{C}_b^e\delta\boldsymbol{\omega}_{ib}^b\end{aligned} \tag{11.43}$$

Substituting from the above into Eq. (11.41) gives

$$\dot{\boldsymbol{\varsigma}} = -\tilde{\boldsymbol{\omega}}_{be}^{e}\boldsymbol{\varsigma} - \tilde{\boldsymbol{\omega}}_{ib}^{e}\boldsymbol{\varsigma} - \mathbf{C}_b^e \delta\boldsymbol{\omega}_{ib}^{b}$$

which upon simplification renders the attitude error equation

$$\dot{\boldsymbol{\varsigma}} = -\tilde{\boldsymbol{\omega}}_{ie}^{e}\boldsymbol{\varsigma} - \mathbf{C}_b^e \delta\boldsymbol{\omega}_{ib}^{b} \tag{11.44}$$

11.3.2 Velocity Error

The velocity error is derived from Eq. (5.23)

$$\delta\dot{\mathbf{v}}^{e} = \delta\mathbf{a}^{e} - 2\boldsymbol{\omega}_{ie}^{e} \times \delta\mathbf{v}^{e} + \delta\mathbf{g}^{e} \tag{11.45}$$

Notice that

$$\begin{aligned} \delta\mathbf{a}^{e} &= \delta\left(\mathbf{C}_b^e \mathbf{a}^{b}\right) \\ &= \left(\delta\mathbf{C}_b^e\right)\mathbf{a}^{b} + \mathbf{C}_b^e \delta\mathbf{a}^{b} \end{aligned} \tag{11.46}$$

Substituting from Eq. (11.40) gives

$$\begin{aligned} \delta\mathbf{a}^{e} &= -\tilde{\boldsymbol{\varsigma}}\mathbf{C}_b^e \mathbf{a}^{b} + \mathbf{C}_b^e \delta\mathbf{a}^{b} \\ &= -\tilde{\boldsymbol{\varsigma}}\mathbf{a}^{e} + \mathbf{C}_b^e \delta\mathbf{a}^{b} \\ &= \tilde{\mathbf{a}}^{e}\boldsymbol{\varsigma} + \mathbf{C}_b^e \delta\mathbf{a}^{b} \end{aligned} \tag{11.47}$$

Substituting from the above in Eq. (11.45) gives the velocity error equation

$$\delta \dot{\mathbf{v}}^e = -2\tilde{\boldsymbol{\omega}}_{ie}^e \delta \mathbf{v}^e + \tilde{\mathbf{a}}^e \varsigma + \mathbf{C}_b^e \delta \mathbf{a}^b + \delta \mathbf{g}^e \tag{11.48}$$

11.3.3 Position Error

The position error is given by

$$\delta \dot{\mathbf{p}}^e = \delta \mathbf{v}^e \tag{11.49}$$

11.3.4 Earth Frame Error State Equation

Ignoring the gravity error as a noise term, the error state equation in the Earth frame coordinates can now be given as

$$\frac{d}{dt}\begin{bmatrix} \delta \mathbf{p}^e \\ \delta \mathbf{v}^e \\ \varsigma \\ \delta \boldsymbol{\omega}_{ib}^b \\ \delta \mathbf{a}^b \end{bmatrix} = \begin{bmatrix} \mathbf{0} & \mathbf{I} & \mathbf{0} & \mathbf{0} & \mathbf{0} \\ \mathbf{0} & -2\tilde{\boldsymbol{\omega}}_{ie}^e & \tilde{\mathbf{a}}^e & \mathbf{0} & \mathbf{C}_b^e \\ \mathbf{0} & \mathbf{0} & -\tilde{\boldsymbol{\omega}}_{ie}^e & -\mathbf{C}_b^e & \mathbf{0} \\ \mathbf{0} & \mathbf{0} & \mathbf{0} & \mathbf{0} & \mathbf{0} \\ \mathbf{0} & \mathbf{0} & \mathbf{0} & \mathbf{0} & \mathbf{0} \end{bmatrix} \begin{bmatrix} \delta \mathbf{p}^e \\ \delta \mathbf{v}^e \\ \varsigma \\ \delta \boldsymbol{\omega}_{ib}^b \\ \delta \mathbf{a}^\mathrm{b} \end{bmatrix} \tag{11.50}$$

11.4 Inertial Sensors Error Models

Equations (11.38) and (11.50) describe the navigation error propagation in the navigation frame and the ECEF frame respectively. However both lack the noise dynamic model for the gyro and accelerometer sensors. The sensor measurements are corrupted by multitude of error sources such as biases, scale factors, orthogonality misalignments [1]. For practical considerations, sensor noise is typically simplified, as shown in Fig. 11.2, by a first order Gauss Markov process. Therefore, the noise models for the gyro and accelerometers errors are

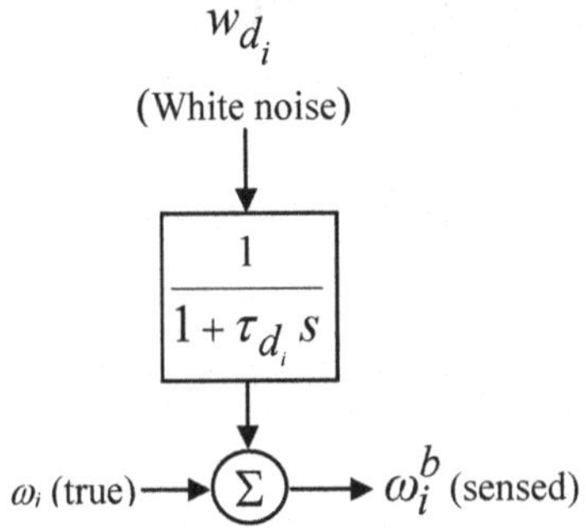

Figure 11.2a Gyro i measurement model

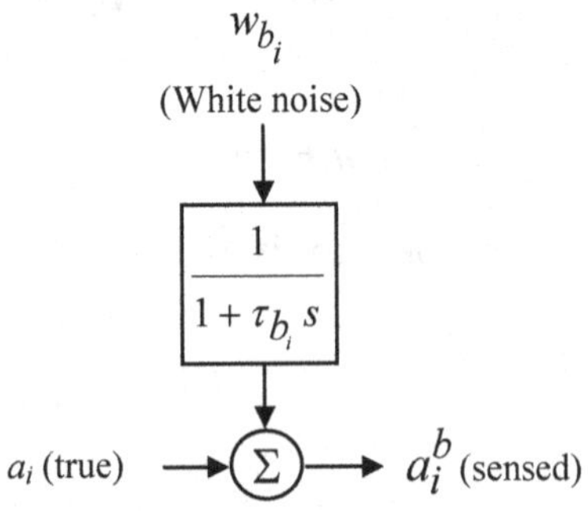

Figure 11.2b Accelerometer i measurement model

$$\frac{d}{dt}\delta\boldsymbol{\omega}_{ib}^{b} = -\mathbf{T}_d^{-1}\delta\boldsymbol{\omega}_{ib}^{b} + \mathbf{T}_d^{-1}\mathbf{w}_d \tag{11.51}$$

$$\frac{d}{dt}\delta\mathbf{a}^{b} = -\mathbf{T}_b^{-1}\delta\mathbf{a}^{b} + \mathbf{T}_b^{-1}\mathbf{w}_b \tag{11.52}$$

where $\mathbf{T}_d$ and $\mathbf{T}_b$ are diagonal matrices that comprise the correlation time constants, and $\mathbf{w}_d$ and $\mathbf{w}_b$ are two white noise processes whose power spectral densities are given by

$$\begin{aligned} \text{PSD}_{gyro}(\mathbf{w}_d) &= \mathbf{Q}_d \ (\text{rad/sec})^2/\text{Hz} \\ \text{PSD}_{accel}(\mathbf{w}_b) &= \mathbf{Q}_b \ (\text{m/sec}^2)^2/\text{Hz} \end{aligned} \tag{11.53}$$

To simplify the error equations, $\mathbf{w}_d$ and $\mathbf{w}_b$ will be normalized with the associated correlation times so that

$$\begin{aligned} \bar{\mathbf{w}}_d &= \mathbf{T}_d^{-1}\mathbf{w}_d \\ \bar{\mathbf{w}}_b &= \mathbf{T}_b^{-1}\mathbf{w}_b \end{aligned} \tag{11.54}$$

With these two equations, the error model in the navigation frame is given by

$$\frac{d}{dt}\begin{bmatrix}\delta\mathbf{p}\\ \delta\mathbf{v}^n\\ \boldsymbol{\eta}\\ \delta\boldsymbol{\omega}_{ib}^b\\ \delta\mathbf{a}^b\end{bmatrix}=\begin{bmatrix}\mathbf{C}_{pp} & \mathbf{C}_{pv} & \mathbf{0} & \mathbf{0} & \mathbf{0}\\ \tilde{\mathbf{v}}^n(\mathbf{C}_{\omega np}+2\mathbf{C}_{\omega ep})+\mathbf{G}_p & (\tilde{\mathbf{v}}^n\mathbf{C}_{\omega nv}-\tilde{\boldsymbol{\omega}}^n) & \tilde{\mathbf{a}}^n & \mathbf{0} & \mathbf{C}_b^n\\ (\mathbf{C}_{\omega np}+\mathbf{C}_{\omega ep}) & \mathbf{C}_{\omega nv} & -\tilde{\boldsymbol{\omega}}_{in}^n & -\mathbf{C}_b^n & \mathbf{0}\\ \mathbf{0} & \mathbf{0} & \mathbf{0} & -\mathbf{T}_d^{-1} & \mathbf{0}\\ \mathbf{0} & \mathbf{0} & \mathbf{0} & \mathbf{0} & -\mathbf{T}_b^{-1}\end{bmatrix}\begin{bmatrix}\delta\mathbf{p}\\ \delta\mathbf{v}^n\\ \boldsymbol{\eta}\\ \delta\boldsymbol{\omega}_{ib}^b\\ \delta\mathbf{a}^b\end{bmatrix}+\begin{bmatrix}\mathbf{0}\\ \mathbf{0}\\ \mathbf{0}\\ \bar{\mathbf{w}}_d\\ \bar{\mathbf{w}}_b\end{bmatrix}$$

(11.55)

and the inertial navigation error equation in the ECEF frame is given by

$$\frac{d}{dt}\begin{bmatrix}\delta\mathbf{p}^e\\ \delta\mathbf{v}^e\\ \varsigma\\ \delta\boldsymbol{\omega}_{ib}^b\\ \delta\mathbf{a}^b\end{bmatrix}=\begin{bmatrix}\mathbf{0} & \mathbf{I} & \mathbf{0} & \mathbf{0} & \mathbf{0}\\ \mathbf{0} & -2\tilde{\boldsymbol{\omega}}_{ie}^e & \tilde{\mathbf{a}}^e & \mathbf{0} & \mathbf{C}_b^e\\ \mathbf{0} & \mathbf{0} & -\tilde{\boldsymbol{\omega}}_{ie}^e & -\mathbf{C}_b^e & \mathbf{0}\\ \mathbf{0} & \mathbf{0} & \mathbf{0} & -\mathbf{T}_d^{-1} & \mathbf{0}\\ \mathbf{0} & \mathbf{0} & \mathbf{0} & \mathbf{0} & -\mathbf{T}_b^{-1}\end{bmatrix}\begin{bmatrix}\delta\mathbf{p}^e\\ \delta\mathbf{v}^e\\ \varsigma\\ \delta\boldsymbol{\omega}_{ib}^b\\ \delta\mathbf{a}^b\end{bmatrix}+\begin{bmatrix}\mathbf{0}\\ \mathbf{0}\\ \mathbf{0}\\ \bar{\mathbf{w}}_d\\ \bar{\mathbf{w}}_b\end{bmatrix}$$

(11.56)

To implement the Kalman filter equations, we need to convert the error models given in the above equations – which are in the continuous time domain –to the discrete time domain and determine the process noise matrix associated with this model. In the following we digress on the theoretical aspects for achieving our objectives. It is noticed that Eqs. (11.55) and (11.56) are modeled after the continuous time equation

$$\dot{\mathbf{x}}=\mathbf{A}(t)\mathbf{x}+\mathbf{w}(t) \tag{11.57}$$

Assuming that the input $\mathbf{w}(t)$ is white noise process then its covariance takes the form of

$$E\,\mathbf{w}(t)\mathbf{w}'(s) = \mathbf{Q}\delta(t-s) \tag{11.58}$$

where δ is the dirac delta and should not be confused with the variation operator used in the earlier sections. Equation (11.57) may be converted to the discrete time recursive form of

$$\mathbf{x}_{n+1} = \mathbf{F}_n\mathbf{x}_n + \mathbf{u}_n \tag{11.59}$$

The notation in the above equation implies that, $\mathbf{x}_n = \mathbf{x}(t_n) = \mathbf{x}\ (n\Delta T)$, where ΔT is the computer integration time interval. When **A** is a constant matrix then

$$\mathbf{F}_n = \exp(\mathbf{A}\Delta T) \tag{11.60}$$

$$\mathbf{u}_n = \int_0^{\Delta T} \exp(A\tau)w(t_{n+1}-\tau)d\tau \tag{11.61}$$

If **A** is not constant then for sufficiently small time interval, Eq. (11.60) may be approximated with

$$\mathbf{F}_n = \mathbf{I} + \mathbf{A}(t_n)\Delta T \tag{11.62}$$

The process covariance matrix for the discrete time process, $\mathbf{u}_n$, is

$$\begin{aligned}\mathbf{Q}_n &= E\mathbf{u}_n\mathbf{u}'_n \\ &= E\int_0^{\Delta T}\int_0^{\Delta T} \exp(\tau\mathbf{A})E\mathbf{w}(t_{n+1}-\tau)\mathbf{w}'(t_{n+1}-\eta)\exp(\mathbf{A}'\eta)\,d\tau\,d\eta\end{aligned} \tag{11.63}$$

Substituting from Eq. (11.58) in the above

$$\mathbf{Q}_n = \int_0^{\Delta T}\int_0^{\Delta T} \exp(\tau\mathbf{A})\mathbf{Q}\delta(\tau-\eta)\exp(\eta\mathbf{A}')\,d\tau\,d\eta$$
$$= \int_0^{\Delta T} \exp(\tau\mathbf{A})\mathbf{Q}\exp(\tau\mathbf{A}')d\tau \tag{11.64}$$

When ΔT is small the above integral can be approximated by

$$\mathbf{Q}_n = \mathbf{Q}\Delta T \tag{11.65}$$

Yet a better approximation, using the mid-point rule [2], is given by

$$\mathbf{Q}_n = \exp(\tfrac{\Delta T}{2}\mathbf{A})\mathbf{Q}\Delta T\exp(\tfrac{\Delta T}{2}\mathbf{A}') \tag{11.66}$$

The process noise matrix, $\mathbf{Q}$, in the above equation is given by

$$\mathbf{Q} = \text{diag}(\mathbf{0}_{3\times3}, \mathbf{0}_{3\times3}, \mathbf{0}_{3\times3}, \mathbf{Q}_d, \mathbf{Q}_b) \tag{11.67}$$

where $\mathbf{Q}_d$ and $\mathbf{Q}_b$ are these matrices given in Eq. (11.53). Substituting for $\mathbf{Q}$ from Eq. (11.67) into Eq. (11.65) or Eq. (11.66) we obtain the desired discrete process covariance matrix. Notice that $\mathbf{A}$ in the above equation denotes the state coefficient matrix in Eqs. (11.55) or (11.56).

11.5 The Global Positioning System

For our purposes, the GPS is a constellation of 24 satellites that orbit the Earth in 6 different orbits, with 4 satellites in each orbit. The GPS is so designed that at any point on Earth and at any instant, there are at least four visible satellites in the sky. On specific frequencies, each satellite transmits signals that reveal its position in the sky and the time at which this data is taken. A GPS receiver tuned to the transmitting frequencies can measure the time delay of the satellite transmitted signal and consequently converts it to a distance since these signals travel at the speed of light. In fact a multiple channel receiver can simultaneously track multiple satellite signals and compute the simultaneous time delays of each satellite signal. Ideally, with a processor on board, a set of three satellites should enable the receiver to compute its position in this coordinate system. However because the clock on board the receiver is almost always different than the GPS timing there is a timing bias that must be accounted for. For that the measured range between the receiver and satellite is called the pseudo range and therefore a minimum of four satellites is needed to compute the position.

It should be noted that all the positioning raw data in the GPS are computed with respect to a three-dimensional Earth Centered Earth fixed (ECEF) Cartesian frame. Now if with respect to the ECEF, the ith satellite position is (x_i, y_i, z_i) the receiver position is (x, y, z), the receiver clock bias is Δ, the speed of light is c, and the receiver range measurement to this satellite is ρ_i then these variables will be related by the equation

$$(x-x_i)^2+(y-y_i)^2+(z-z_i)^2=(\rho_i-c\Delta)^2, \quad i=1,2,3,4 \tag{11.68}$$

The above can be expressed in terms of the following equations

$$\rho_i = r_i + b \tag{11.69}$$

$$b = c\Delta \tag{11.70}$$

$$r_i=\sqrt{(x-x_i)^2+(y-y_i)^2+(z-z_i)^2} \tag{11.71}$$

where b is the clock bias normalized by the speed of light and r_i is the true range of the receiver relative to the satellite. Equation (11.71) could be solved iteratively by linearization as follows. Let

$$\mathbf{H} = \begin{bmatrix} \frac{\partial\rho_1}{\partial x} & \frac{\partial\rho_1}{\partial y} & \frac{\partial\rho_1}{\partial z} & \frac{\partial\rho_1}{\partial b} \\ \frac{\partial\rho_2}{\partial x} & \frac{\partial\rho_2}{\partial y} & \frac{\partial\rho_2}{\partial z} & \frac{\partial\rho_2}{\partial b} \\ \frac{\partial\rho_3}{\partial x} & \frac{\partial\rho_3}{\partial y} & \frac{\partial\rho_3}{\partial z} & \frac{\partial\rho_3}{\partial b} \\ \frac{\partial\rho_4}{\partial x} & \frac{\partial\rho_4}{\partial y} & \frac{\partial\rho_4}{\partial z} & \frac{\partial\rho_4}{\partial b} \end{bmatrix} \tag{11.72}$$

Carrying out the partial derivatives and substituting in the above gives

$$\mathbf{H}(x,y,z) = \begin{bmatrix} \frac{(x-x_1)}{r_1} & \frac{(y-y_1)}{r_1} & \frac{(z-z_1)}{r_1} & 1 \\ \frac{(x-x_2)}{r_2} & \frac{(y-y_2)}{r_2} & \frac{(z-z_2)}{r_2} & 1 \\ \frac{(x-x_3)}{r_3} & \frac{(y-y_3)}{r_3} & \frac{(z-z_3)}{r_3} & 1 \\ \frac{(x-x_4)}{r_4} & \frac{(y-y_4)}{r_4} & \frac{(z-z_4)}{r_4} & 1 \end{bmatrix} \tag{11.73}$$

At the kth iteration step, the solution is denoted by $(x, y, z, b)^k$, thus

$$\begin{bmatrix} x \\ y \\ z \\ b \end{bmatrix}^{k+1} = \begin{bmatrix} x \\ y \\ z \\ b \end{bmatrix}^{k} + \mathbf{H}^{-1}(x^k, y^k, z^k) \begin{bmatrix} \rho_1 - r_1(x^k, y^k, z^k) - b^k \\ \rho_2 - r_2(x^k, y^k, z^k) - b^k \\ \rho_3 - r_3(x^k, y^k, z^k) - b^k \\ \rho_4 - r_4(x^k, y^k, z^k) - b^k \end{bmatrix} \tag{11.74}$$

The position vector could be initialized with the zero vector which is equivalent to be at the center of the Earth.

Remarks:

1. As mentioned earlier, we will address only the loosely coupled approach for interfacing the GPS to the INS. In this approach the GPS and INS data processing are almost completely decoupled. To implement this approach, the GPS receiver provides the craft position (x,y,z) in the ECEF frame.
2. Depending on the mechanization of the Kalman filter, the ECEF coordinates provided by the GPS may need to be converted into the Nav-frame coordinates (λ, ϕ, h).
3. The pseudo range measurement in Eq. (11.69) not only is affected by the clock bias but also with other error terms such as atmospheric delays, multipath and random noise [1].

In the following we shall determine the relations between the coordinates in the Nav-frame to those in the ECEF. Let p be a point with coordinates in the Nav-frame given by (λ, ϕ, h), the longitude, latitude and normal elevation respectively.

From Fig. 11.3 and Eqs. (D.21) and (D.22), the horizontal and vertical projections, p_h and p_v, of point p in the meridian plane, in which it locates, are given by

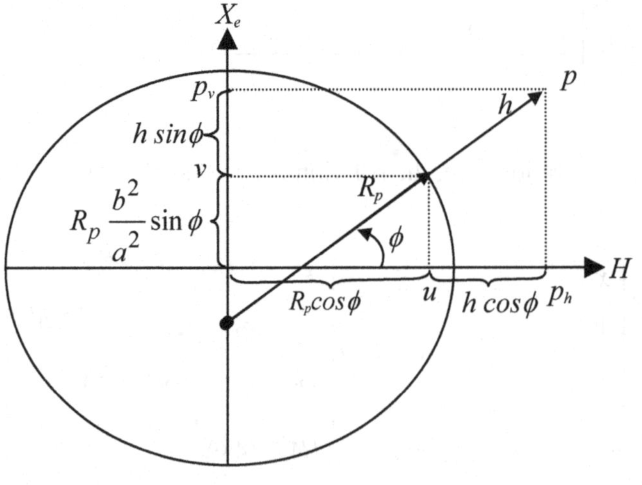

Figure 11.3 Geometry in the Meridian Plane

$$p_h = (R_p + h)\cos\phi$$

$$p_v = (R_p \frac{b^2}{a^2} + h)\sin\phi$$

Using the Earth axes notations in Fig. 4.4, and analyzing p_h along the principal meridian plane through the longitudinal angle λ, we get

$$z_e = -p_h \cos\lambda = -(R_p + h)\cos\phi\cos\lambda \tag{11.75}$$

$$y_e = p_h \sin\lambda = (R_p + h)\cos\varphi\sin\lambda \tag{11.76}$$

$$x_e = (R_p \frac{b^2}{a^2} + h)\sin\phi \tag{11.77}$$

where (x_e, y_e, z_e) are the coordinates of the receiver in the ECEF frame.

11.6 Mechanization of the INS/GPS Equations

We shall provide two mechanizations to interface the INS and the GPS with a Kalman filter, see Appendix H. The first mechanization uses the navigation frame position coordinates (ϕ, λ, h), see Fig. 11.4a. This mechanization uses the inertial error state equation, in the Nav-frame, as provided in Eq. (11.55). In this case the ECEF-frame GPS solution should be converted via Eqs. (11.75)-(11.77) to Nav-frame solution, i.e. in terms of (ϕ, λ, h).

The other mechanization uses the ECEF frame position coordinates (x_e, y_e, z_e) as in Fig. 11.4b. This mechanization uses the inertial error state equation in the ECEF frame as provided in Eq. (11.56). Here the ECEF-frame GPS solution remains unchanged.

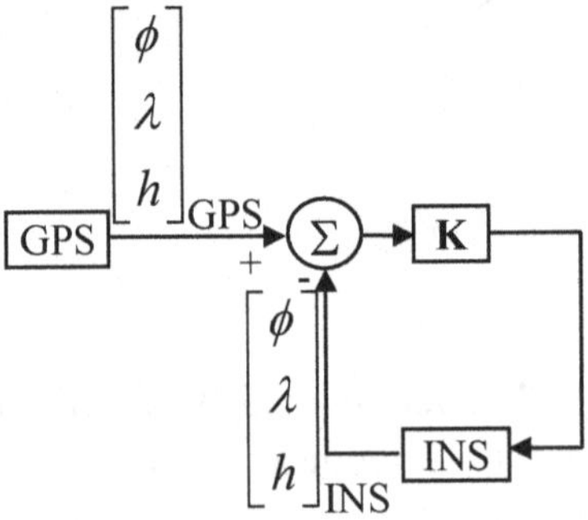

Figure 11.4a Navigation Frame Mechanization

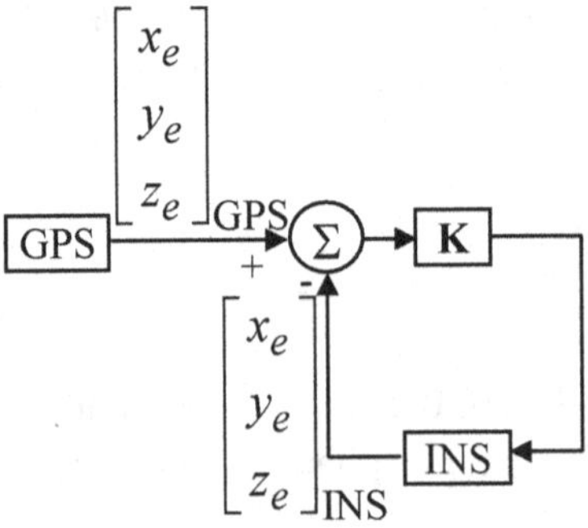

Figure 11.4b ECEF Frame Mechanization

Implementing the Kalman filter equations requires the determination of these parameters:

1. For the physical process: the state transition matrix, $\mathbf{F}_n$, and the process noise covariance matrix, $\mathbf{Q}_n$.
2. For the observability process: the measurement matrix, $\mathbf{H}_n$, and the measurement covariance matrix, $\mathbf{R}_n$.
3. An initial estimate of the state, $\mathbf{x}_0$, and the associated error covariance matrix, $\mathbf{P}_0$.

The state transition matrix, $\mathbf{F}_n$, is obtained by substituting the state coefficient matrix from Eq. (11.55) or Eq. (11.56), depending on the mechanization, into Eq. (11.60) or Eq. (11.62), depending on the desired accuracy. From the inertial sensor specifications, the noise covariance matrix, $\mathbf{Q}_n$, in Eq. (11.67) is determined. Then by substituting in Eq. (11.65) or Eq. (11.66), depending on the desired accuracy, we get the process noise matrix in the discrete time domain.

In both mechanizations, the measurement matrix is given by

$$\mathbf{H}_n = \begin{bmatrix} \mathbf{I}_{3\times3} & \mathbf{0}_{3\times3} & \mathbf{0}_{3\times3} & \mathbf{0}_{3\times3} & \mathbf{0}_{3\times3} \end{bmatrix} \qquad (11.78)$$

The measurement noise matrix, $\mathbf{R}_n$, could be passed from the error covariance matrix of the GPS Kalman filter. This is another Kalman filter to estimate the craft position and accessible only by the GPS data.

Independent of the inertial data, the GPS receiver could be processing the satellite data using a Kalman filter of its own

References

1. J. A. Farrell & M. Barth, The Global Positioning System & Inertial Navigation, McGraw-Hill, New York, 1999.
2. http://www.math.rutgers.edu/~rcostin/152/152-formulas.pdf (midpoint rule)

Appendix A

The Vector Dot and Cross Products

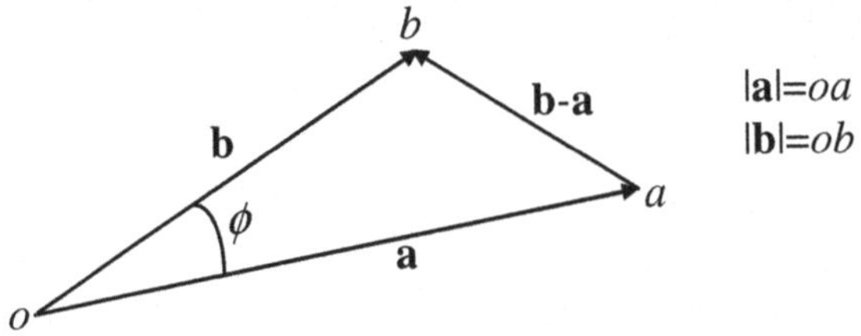

Figure A.1

Figure A.1 depicts two coplanar vectors **a** and **b** coinciding at point o and in their co-plane lies the triangle oab. It is desired to show how the cosine of the (physical) angle between **a** and **b** can described by their vector dot product. The cosine law implies that

$$|\mathbf{b}-\mathbf{a}|^2 = |\mathbf{a}|^2 + |\mathbf{b}|^2 - 2|\mathbf{a}||\mathbf{b}|\cos\phi \Rightarrow$$

$$\cos\phi = \frac{|\mathbf{a}|^2 + |\mathbf{b}|^2 - |\mathbf{b}-\mathbf{a}|^2}{2|\mathbf{a}||\mathbf{b}|}$$

Substituting for the lengths in terms of the respective dot products we get

$$\begin{aligned}\cos\phi &= \frac{\mathbf{a}\cdot\mathbf{a} + \mathbf{b}\cdot\mathbf{b} - (\mathbf{b}-\mathbf{a})\cdot(\mathbf{b}-\mathbf{a})}{2|\mathbf{a}||\mathbf{b}|} \\ &= \frac{(\mathbf{a}\cdot\mathbf{b})}{|\mathbf{a}||\mathbf{b}|}\end{aligned} \tag{A.1}$$

where $(\mathbf{a}\cdot\mathbf{b})$ is the vector dot product of **a** and **b**.

Now we explore the cross product property of **a** and **b**. Suppose that we have a three dimensional space defined by the orthonormal vectors $(\mathbf{e}_1,\mathbf{e}_2,\mathbf{e}_3)$ and in which the two unit vectors **a** and **b** lie in the plane of the vectors $(\mathbf{e}_1,\mathbf{e}_2)$. With no loss of generality we shall assume that **a** is along $\mathbf{e}_1$.

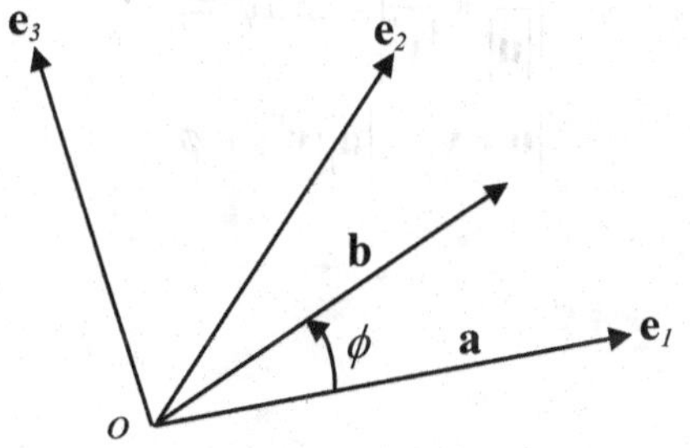

Figure A.2

In this space the vectors **a** and **b** are represented by

$$\mathbf{a} = \begin{pmatrix}1 & 0 & 0\end{pmatrix}$$

$$\mathbf{b} = \begin{pmatrix}\cos\phi & \sin\phi & 0\end{pmatrix}$$

From Eq. (1.11), **c**, the cross product of **a** and **b** is given by

$$\begin{aligned}\mathbf{c} &= \mathbf{a}\times\mathbf{b}\\ &= \begin{pmatrix}0 & 0 & \sin\phi\end{pmatrix}\\ &= \sin\phi\begin{pmatrix}0 & 0 & 1\end{pmatrix}\end{aligned}$$

It can be seen that **c** is orthogonal to **a** and **b**, either by verifying that the inner product of **c** with both **a** and **b** is zero, or by observing that **c** is along $\mathbf{e}_3$ which is orthogonal to the plane of the vectors $\mathbf{e}_1$ and $\mathbf{e}_2$. Notice also that the magnitude of **c** – the cross product vector of the unit vectors **a** and **b** – is the sine of the angle between them.

$$|\mathbf{a} \times \mathbf{b}| = \sin \phi$$

If **u** and **v** are two arbitrary (non unit) vectors for which ϕ is the angle between them, then we can substitute in the above equation to get

$$\left| \frac{\mathbf{u}}{|\mathbf{u}|} \times \frac{\mathbf{v}}{|\mathbf{v}|} \right| = \sin \phi \Rightarrow$$
$$|\mathbf{u} \times \mathbf{v}| = |\mathbf{u}||\mathbf{v}| \sin \phi \tag{A.2}$$

Appendix B

Introduction to Quaternion Algebra

Quaternions, called hyper complex numbers, are generalization of complex numbers, and were originally conceived by R.W. Hamilton in 1843 [1]. A complex number is given by

$$c = a + ib \tag{B.1}$$

where a and b are both real numbers and i is the imaginary number which is given by

$$i = \sqrt{-1} \tag{B.2}$$

As evident, a complex number is the sum of two elements one real and one imaginary. Complex numbers may be thought of as vectors in a two dimensional space with two element basis $\{e_0, e_1\}$, where e_0=1 and e_1=i. As such, c in Eq. (B.1) can be represented by

$$c = a\, e_0 + b\, e_1 \tag{B.3}$$

A quaternion, made of four elements, is a generalization of a complex number. Similar to a complex vectors, they can be represented in a four dimensional space using four element basis $\{\mathbf{e}_0, \mathbf{e}_1, \mathbf{e}_2, \mathbf{e}_3\}$ defined by

$$\mathbf{e}_0 = \begin{bmatrix} 1 & 0 \\ 0 & 1 \end{bmatrix} \tag{B.4a}$$

$$\mathbf{e}_1 = \begin{bmatrix} i & 0 \\ 0 & -i \end{bmatrix} \tag{B.4b}$$

$$\mathbf{e}_2 = \begin{bmatrix} 0 & i \\ i & 0 \end{bmatrix} \tag{B.4c}$$

$$\mathbf{e}_3 = \begin{bmatrix} 0 & -1 \\ 1 & 0 \end{bmatrix} \tag{B.4d}$$

Hamilton's original description of this basis was given by [2]

$$\mathbf{e}_1\mathbf{e}_1 = \mathbf{e}_2\mathbf{e}_2 = \mathbf{e}_3\mathbf{e}_3 = \mathbf{e}_1\mathbf{e}_2\mathbf{e}_3 = -\mathbf{e}_0 \tag{B.5}$$

From which we could deduce that

$$\begin{aligned} &\mathbf{e}_1\mathbf{e}_2 = \mathbf{e}_3, \quad \mathbf{e}_2\mathbf{e}_3 = \mathbf{e}_1, \quad \mathbf{e}_3\mathbf{e}_1 = \mathbf{e}_2, \\ &\mathbf{e}_2\mathbf{e}_1 = -\mathbf{e}_3, \quad \mathbf{e}_3\mathbf{e}_2 = -\mathbf{e}_1, \quad \mathbf{e}_1\mathbf{e}_3 = -\mathbf{e}_2. \end{aligned} \tag{B.6}$$

With the basis $\{\mathbf{e}_0, \mathbf{e}_1, \mathbf{e}_2, \mathbf{e}_3\}$, a quaternion $\mathbf{Q}$ is represented by

$$\mathbf{Q} = q_0\mathbf{e}_0 + q_1\mathbf{e}_1 + q_2\mathbf{e}_2 + q_3\mathbf{e}_3 \tag{B.7}$$

where (q_0, q_1, q_2, q_3) are scalars. When we substitute for the elements of the basis from Eq. (B.4) in the above equation, the quaternion becomes

$$\mathbf{Q} = q_0\mathbf{e}_0 + q_1\mathbf{e}_1 + q_2\mathbf{e}_2 + q_3\mathbf{e}_3$$
$$= q_0\begin{bmatrix}1 & 0\\ 0 & 1\end{bmatrix} + q_1\begin{bmatrix}i & 0\\ 0 & -i\end{bmatrix} + q_2\begin{bmatrix}0 & i\\ i & 0\end{bmatrix} + q_3\begin{bmatrix}0 & -1\\ 1 & 0\end{bmatrix} \quad \text{(B.8)}$$
$$= \begin{bmatrix}q_0 + iq_1 & -q_3 + iq_2\\ q_3 + iq_2 & q_0 - iq_1\end{bmatrix}$$

Even though Eq. (B.8) is the formal representation of a quaternion, we could present it as a four dimensional vector for computational purposes. For derivation and for short hand the quaternion is often represented by a scalar and a three dimensional vector. Accordingly, the quaternion in Eq. (B.8) would be represented by $\mathbf{Q} = \{q_0, \mathbf{q}\}$, where $\mathbf{q} = [q_1 \quad q_2 \quad q_3]$. The physical significance of the scalar and vector components of the quaternion has been discussed in Chapter 2.

Let us define an arbitrary quaternion **P** by

$$\mathbf{P} = p_0\mathbf{e}_0 + p_1\mathbf{e}_1 + p_2\mathbf{e}_2 + p_3\mathbf{e}_3 \quad \text{(B.9)}$$

For arbitrary scalars *a* and *b*, the linear combination of **P** and **Q**

$$\begin{aligned}\mathbf{S} &= a\mathbf{P} + b\mathbf{Q}\\ &= a\left(p_0\mathbf{e}_0 + p_1\mathbf{e}_1 + p_2\mathbf{e}_2 + p_3\mathbf{e}_3\right)\\ &\quad + b\left(q_0\mathbf{e}_0 + q_1\mathbf{e}_1 + q_2\mathbf{e}_2 + q_3\mathbf{e}_3\right)\\ &= \left(ap_0 + bq_0\right)\mathbf{e}_0 + \left(ap_1 + bq_1\right)\mathbf{e}_1\\ &\quad + \left(ap_2 + bq_2\right)\mathbf{e}_2 + \left(ap_3 + bq_3\right)\mathbf{e}_3\end{aligned} \quad \text{(B.10)}$$

is also a quaternion and hence the quaternion is a linear space for all real a and b. In short hand notation the quaternion **P** is given by $\mathbf{P} = \{p_0, \mathbf{p}\}$ where $\mathbf{p} = [p_1 \quad p_2 \quad p_3]$ and the linear sum **S** would be given by

$$\mathbf{S} = \{ap_0 + bq_0, a\mathbf{p} + b\mathbf{q}\}$$

The product of a two quaternions is also a quaternion which can be obtained by carrying out the 2×2 matrix product and utilizing Eq. (B.6). For example, **R** the product of **P** and **Q** is a quaternion given by

$$\begin{aligned}\mathbf{R} &= \mathbf{PQ}\\ &= (p_0\mathbf{e}_0 + p_1\mathbf{e}_1 + p_2\mathbf{e}_2 + p_3\mathbf{e}_3)(q_0\mathbf{e}_0 + q_1\mathbf{e}_1 + q_2\mathbf{e}_2 + q_3\mathbf{e}_3)\\ &= (p_0q_0 - p_1q_1 - p_2q_2 - p_3q_3)\mathbf{e}_0 + (p_1q_0 + p_0q_1 + p_2q_3 - p_3q_2)\mathbf{e}_1\\ &\quad + (p_0q_2 - p_1q_3 + p_3q_1 + p_2q_0)\mathbf{e}_2 + (p_0q_3 + p_1q_2 + p_3q_0 - p_2q_1)\mathbf{e}_3\end{aligned} \tag{B.11}$$

If the above quaternion is given by the short hand $\mathbf{R} = \{r_0, \mathbf{r}\}$ where $\mathbf{r} = [r_1 \quad r_2 \quad r_3]$, then from Eq. (B.11) its four elements are

$$r_0 = p_0q_0 - p_1q_1 - p_2q_2 - p_3q_3 \tag{B.12a}$$

$$\mathbf{r} = \begin{bmatrix} r_1 \\ r_2 \\ r_3 \end{bmatrix} = \begin{bmatrix} p_0q_1 + p_1q_0 + p_2q_3 - p_3q_2 \\ p_0q_2 + p_2q_0 + p_3q_1 - p_1q_3 \\ p_0q_3 + p_3q_0 + p_1q_2 - p_2q_1 \end{bmatrix} \tag{B.12b}$$

From Eq. (B.12) we see that

$$r_0 = p_0q_0 - (\mathbf{p} \cdot \mathbf{q}) \tag{B.13a}$$

$$\begin{aligned}\mathbf{r} &= \begin{bmatrix} p_0q_1 \\ p_0q_2 \\ p_0q_3 \end{bmatrix} + \begin{bmatrix} p_1q_0 \\ p_2q_0 \\ p_3q_0 \end{bmatrix} + \begin{bmatrix} p_2q_3 - p_3q_2 \\ p_3q_1 - p_1q_3 \\ p_1q_2 - p_2q_1 \end{bmatrix}\\ &= p_0\mathbf{q} + q_0\mathbf{p} + \mathbf{p} \times \mathbf{q}\end{aligned} \tag{B.13b}$$

Alternatively we could have performed the quaternion product by carrying out the 2×2 quaternion product as follows:

$$\mathbf{R} = \mathbf{PQ}$$

$$= \begin{bmatrix} p_0 + ip_1 & -p_3 + ip_2 \\ p_3 + ip_2 & p_0 - ip_1 \end{bmatrix} \begin{bmatrix} q_0 + iq_1 & -q_3 + iq_2 \\ q_3 + iq_2 & q_0 - iq_1 \end{bmatrix}$$

$$= \begin{bmatrix} \begin{pmatrix} (p_0q_0 - p_1q_1 - p_2q_2 - p_3q_3) \\ +i(p_1q_0 + p_0q_1 + p_2q_3 - p_3q_2) \end{pmatrix} & \begin{pmatrix} -(p_0q_3 + p_1q_2 + p_3q_0 - p_2q_1) \\ +i(p_0q_2 - p_1q_3 + p_3q_1 + p_2q_0) \end{pmatrix} \\ \begin{pmatrix} (p_0q_3 + p_1q_2 + p_3q_0 - p_2q_1) \\ +i(p_0q_2 - p_1q_3 + p_3q_1 + p_2q_0) \end{pmatrix} & \begin{pmatrix} (p_0q_0 - p_1q_1 - p_2q_2 - p_3q_3) \\ -i(p_1q_0 + p_0q_1 + p_2q_3 - p_3q_2) \end{pmatrix} \end{bmatrix}$$

from which we would obtain the four quaternion elements of **R** exactly as given in Eq. (B.12). From Eq. (B.13) it is observed that the quaternion product is not commutative, that is $\mathbf{PQ} \neq \mathbf{QP}$

The quaternion

$$\mathbf{Q_I} = 1\mathbf{e}_0 + 0\mathbf{e}_1 + 0\mathbf{e}_2 + 0\mathbf{e}_3 = \mathbf{e}_0$$

plays the role of the identity quaternion, as one can see for an arbitrary quaternion **P** that $\mathbf{Q_I P} = \mathbf{PQ_I} = \mathbf{P}$. One can also see that the identity quaternion in short notation is given by $\mathbf{Q_I} = \{1, \mathbf{0}\} = [1,0,0,0]$

Finally, the norm of quaternion **Q**, given by Eq. (B.7), is defined by

$$\text{norm}(\mathbf{Q}) = \sqrt{q_0^2 + q_1^2 + q_2^2 + q_3^2} \tag{B.14}$$

As much as a complex number describes planar rotations (i.e. rotations in two dimensional spaces), a quaternion plays a central role in describing rotations in three dimensional spaces. Quaternions could be found to be a very useful tool not only in developing navigation equations but also in aircraft simulations [3].

References

1. http://mathworld.wolfram.com/Quaternion.html
2. http://www.plus.maths.org/issue32/features/baez/
3. W. F. Phillips, Mechanics of Flight, John Wiley & Sons, New York, 2004.

Appendix C

Simulink® Models

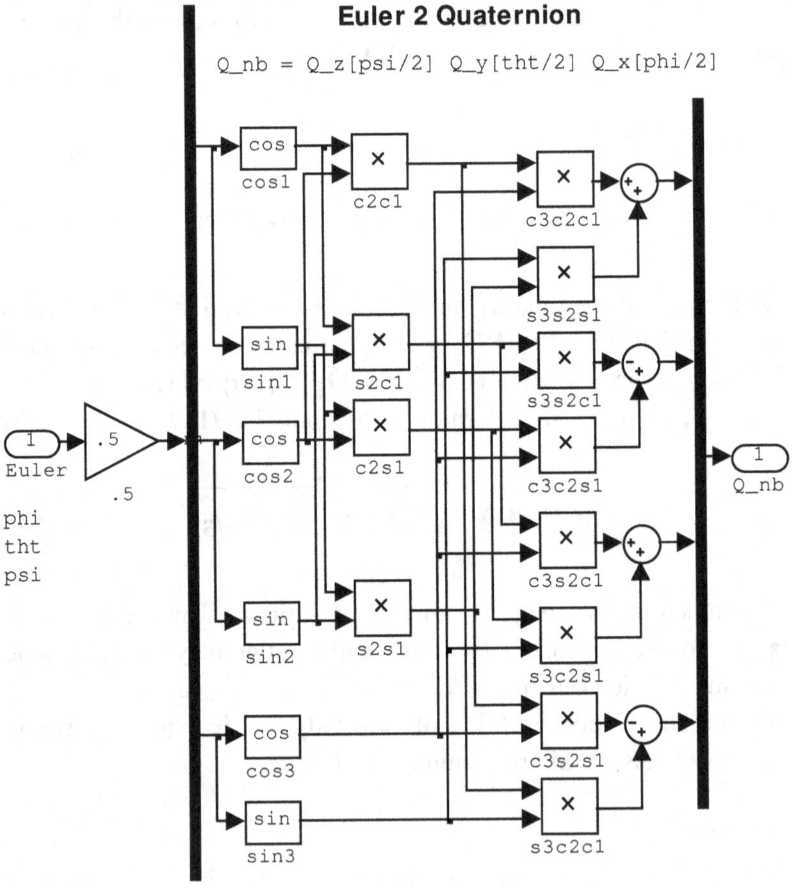

Figure C.1 Euler to Quaternion Model

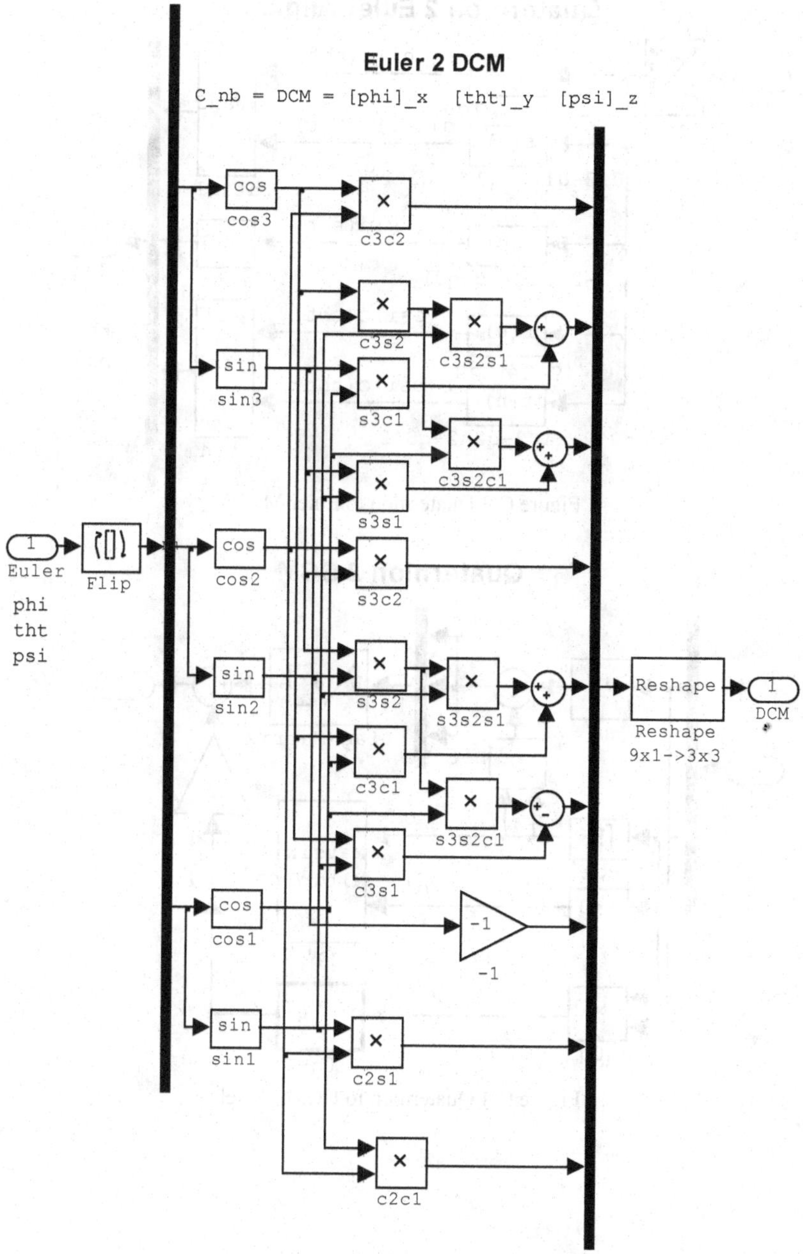

Figure C.2 Euler to DCM Model

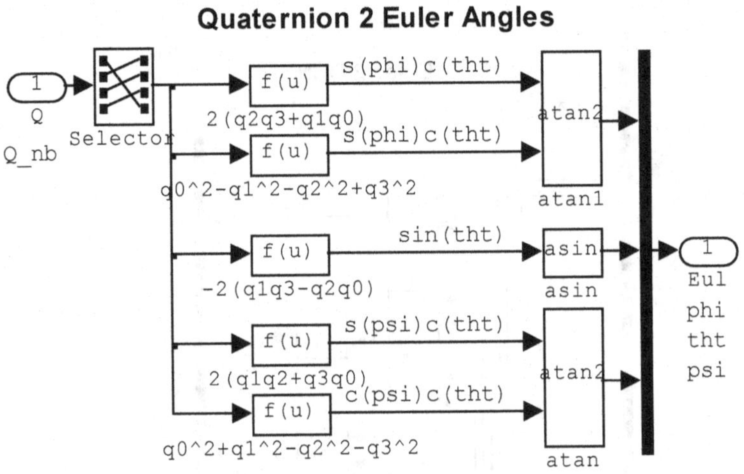

Figure C.3 Quaternion to Euler Model

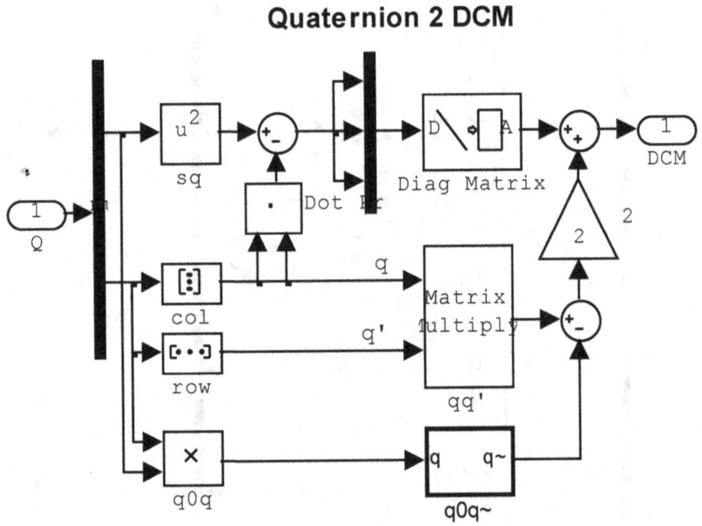

Figure C.4 Quaternion to DCM Model

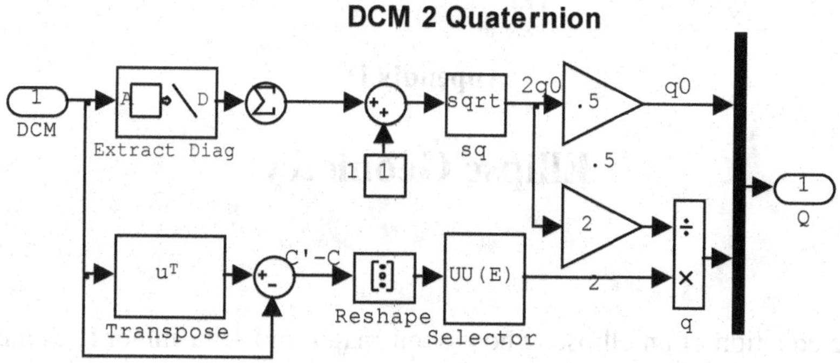

Figure C.5 DCM to Quaternion Model

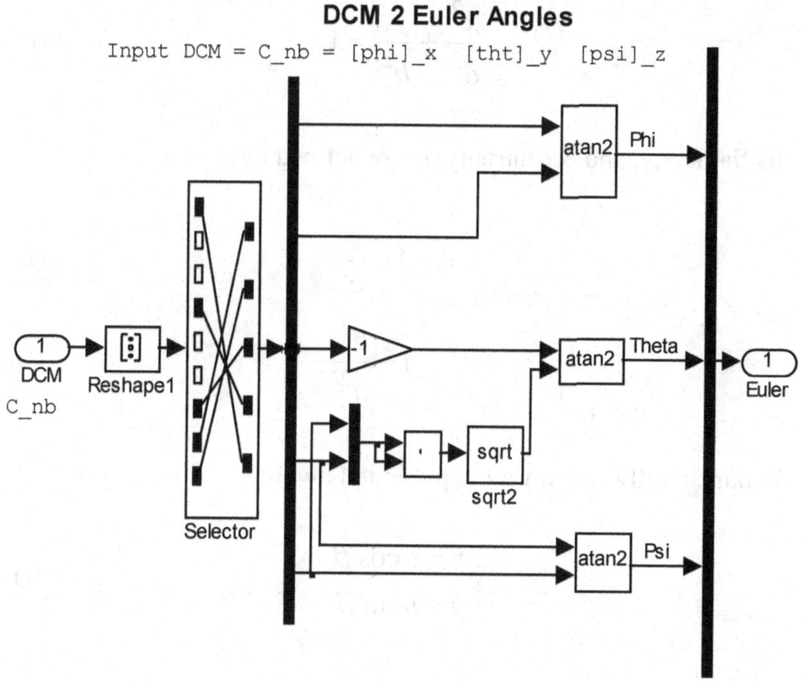

Figure C.6 DCM to Euler Model

Appendix D

Ellipse Geometry

The equation of an ellipse, whose semi-major and semi-minor axes are a and b respectively, is given by

$$\frac{x^2}{a^2}+\frac{y^2}{b^2}=1 \tag{D.1}$$

Its flatness, f, and eccentricity, e, are defined by

$$f=1-\frac{b}{a} \tag{D.2}$$

$$e^2=1-\frac{b^2}{a^2} \tag{D.3}$$

Parametrically, x and y are represented by

$$\begin{aligned} x &= a\cos\beta \\ y &= b\sin\beta \end{aligned} \tag{D.4}$$

The tangent at an arbitrary point is given by

$$\tan\psi=\frac{dy}{dx}=\frac{dy/d\beta}{dx/d\beta}=-\frac{b\cos\beta}{a\sin\beta}=-\frac{b}{a}\cot\beta \tag{D.5}$$

The angles β and ψ are depicted in Fig. D.1. If the orthogonal to the tangent intersects the x-axis at angle ϕ, then $\psi = \frac{\pi}{2} + \phi$ and Eq. (D.5) becomes

$$\tan\beta = -\frac{b}{a}\cot\psi = \frac{b}{a}\tan\phi \tag{D.6}$$

This implies

$$\sec^2\beta = 1 + \tan^2\beta = 1 + \frac{b^2}{a^2}\tan^2\phi = \sec^2\phi\left(\cos^2\phi + \frac{b^2}{a^2}\sin^2\phi\right)$$

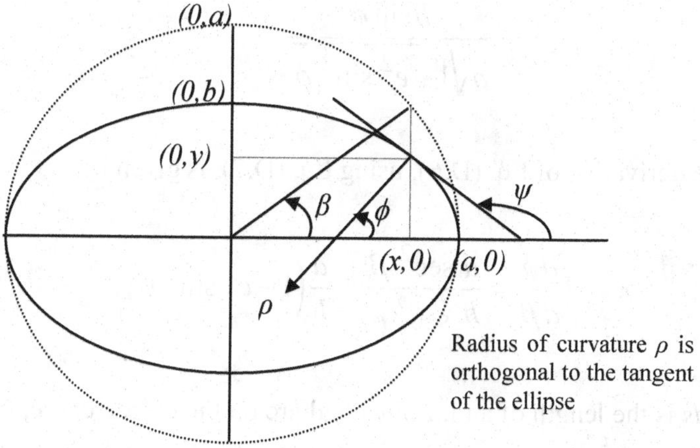

Figure D.1 Ellipse Geometry

Using Eq. (D.3), the above equation can be formed into

$$\sec^2\beta = \sec^2\phi\left(1 - e^2\sin^2\phi\right) \tag{D.7}$$

Taking the square root of both sides of Eq. (D.7) yields

$$\sec\beta = \sec\phi\sqrt{1-e^2\sin^2\phi} \tag{D.8}$$

whose inverse is given by

$$\cos\beta = \frac{\cos\phi}{\sqrt{1-e^2\sin^2\phi}} \tag{D.9}$$

Equations (D.6) and (D.9) imply that

$$\begin{aligned}\sin\beta = \tan\beta\cos\beta &= \frac{b}{a}\frac{\tan\phi\cos\phi}{\sqrt{1-e^2\sin^2\phi}} \\ &= \frac{b\sin\phi}{a\sqrt{1-e^2\sin^2\phi}}\end{aligned} \tag{D.10}$$

The derivative of Eq. (D.6), using Eq. (D.7), is given by

$$\frac{d\phi}{d\beta} = \frac{a}{b}\frac{\sec^2\beta}{\sec^2\phi} = \frac{a}{b}\left(1-e^2\sin^2\phi\right) \tag{D.11}$$

If *ds* is the length of an infinitesimal arc on the ellipse curve, then

$$ds^2 = dx^2 + dy^2 = dy^2[(\frac{dx}{dy})^2 + 1] \tag{D.12}$$

We note from Eq. (D.5) that

$$(\frac{dx}{dy})^2 + 1 = \cot^2\psi + 1 = \tan^2\phi + 1 = \sec^2\phi \tag{D.13}$$

Substituting from Eqs. (D.4) and (D.13) into Eq. (D.12) implies that

$$\left(\frac{ds}{d\beta}\right)^2 = \left(\frac{dy}{d\beta}\right)^2 \left[\left(\frac{dx}{dy}\right)^2 + 1\right] = b^2 \cos^2 \beta \sec^2 \phi = b^2 \frac{\sec^2 \phi}{\sec^2 \beta}$$

Using Eq. (D.7) to eliminate the β term from the above equation gives

$$\left(\frac{ds}{d\beta}\right)^2 = \frac{b^2}{1 - e^2 \sin^2 \phi} \tag{D.14}$$

Since the radius of curvature in the ellipse plane is given by

$$\rho = \frac{ds}{d\psi} \tag{D.15}$$

Substituting from Eqs. (D.11) and (D.14) into Eq. (D.15) yields

$$\begin{aligned} \rho &= \frac{ds/d\beta}{d\psi/d\beta} = \frac{ds/d\beta}{d\phi/d\beta} = \frac{b^2}{a[1 - e^2 \sin^2 \phi]^{3/2}} \\ &= \frac{a(1 - e^2)}{[1 - e^2 \sin^2 \phi]^{3/2}} \end{aligned} \tag{D.16}$$

This radius of curvature is denoted by R_m to distinguish it from other expressions, thus

$$R_m = \frac{a(1 - e^2)}{[1 - e^2 \sin^2 \phi]^{3/2}} \tag{D.17}$$

It should be recognized that Eq. (D.17) determines the radius of curvature for all points on the ellipse curve in the plane in which it lies. We would like to step further and determine the radius of curvature on a revolutionary surface. Of specific interest is an ellipsoidal surface, as it is the Earth's ideal geometry. Figure D.2 depicts a plane ellipse that when revolved about its minor axis – in Earth's case it will be the polar axis – will generate the desired ellipsoid. As point $p(u,v)$ on the ellipse revolves it will prescribe a circle (that will be parallel to the equatorial plane) centered on the axis of rotation and with radius u.

Our objective is to determine the radius of curvature in a plane perpendicular to the ellipse plane at point p. To do that we will let the ellipse plane rotate about the rotation axis an infinitesimal rotation at the end of which point p will be p'. The planes of these two ellipses intersect at the minor axis. By definition, the center of curvature is marked by the intersection of the normal lines at the ends of the arc in this plane. In our case these two normal lines fall in the inclined plane that intersects the minor axis at point c. Hence, pc is the radius of curvature at point p on the inclined plane. It may be of interest to note that the minor axis is the locus of all the centers of curvature for all the inclined planes. Assuming that the inclined plane intersects the major axis (equator) at an angle ϕ, then in case of ϕ=0, the radius of curvature pc will be parallel to the major axis (the equator) and hence the radius of curvature will be u, as we would have expected.

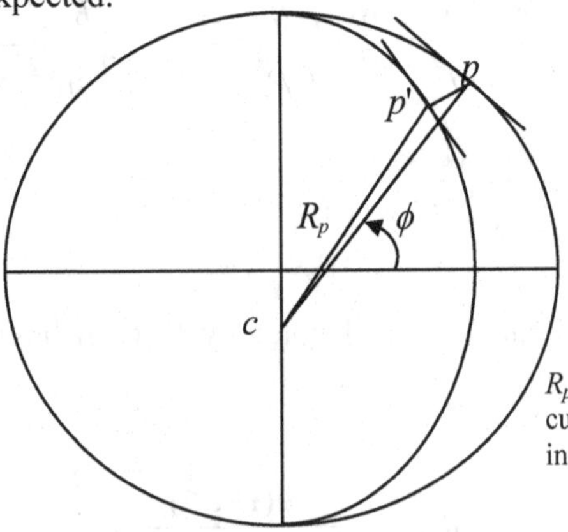

R_p is the radius of curvature in the inclined plane pcp'

Figure D.2 Curvature in Inclined Plane

From Eqs. (D.4), (D.7) and (D.10) if the coordinates of p are (u,v), then

$$u = a\cos\beta = \frac{a\cos\phi}{[1-e^2\sin^2\phi]^{1/2}} \tag{D.18}$$

$$v = b\sin\beta = \frac{b^2}{a}\frac{\sin\phi}{[1-e^2\sin^2\phi]^{1/2}} \tag{D.19}$$

It can be seen in Fig. D.3 that in the inclined plane the radius of curvature is given by

$$R_p = \frac{u}{\cos\phi} = \frac{a}{[1-e^2\sin^2\phi]^{1/2}} \tag{D.20}$$

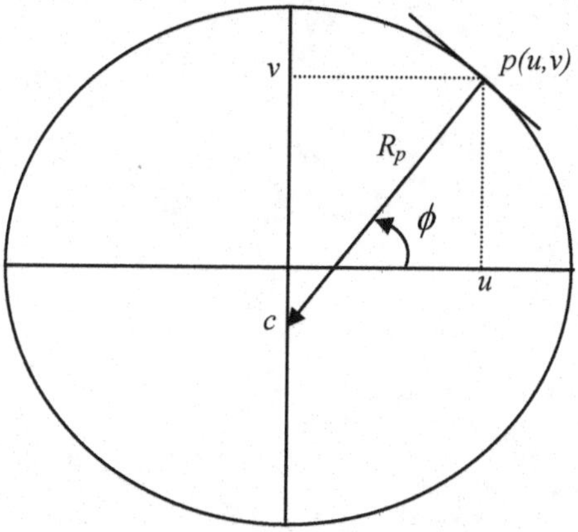

R_p is the radius of curvature in the inclined plane

Figure D.3 Radius of Curvature in Inclined Plane

In terms of R_p and ϕ Eqs. (D.18) and (D.19) may be written as

$$u = R_p\cos\phi \tag{D.21}$$

$$v = \frac{b^2}{a^2} R_p \sin\phi \tag{D.22}$$

References

1. R. Larson, R. Hostetler, B. Edwards, Calculus with Analytic Geometry, D. C. Heath and Company, Lexington, Ma, 1994.
2. D. Varberg, E. Purcell, Calculus with Analytic Geometry, Prentice Hall, Englewood Cliffs, New Jersey, 1992.

Appendix E

Vector Dynamics

In order to apply Newton laws to particles or bodies we must be able to determine their velocities and accelerations. In some cases these kinematics are described in stationary 3-dimensional Cartesian frames and applying the Newton laws is straightforward. However in other cases deriving the kinematics in a stationary frame is not easy. An example is a particle that moves on the surface of a horizontal platform that revolves about its vertical axis. A more pragmatic example is a top spinning on a horizontal surface.

To simplify the analysis we shall study the motion of the frame and the motion of the body and the particle separately. We suppose that a frame rotates about some axis. As it rotates we take two consecutive snap shots of this frame one at times t and $t+\Delta t$ and we denote the frames at these two instances as $\boldsymbol{S}$ and $\boldsymbol{S}'$. In this scenario it is assumed that the two frames share the same origin. In frame $\boldsymbol{S}$, the unit vectors along the (x, y, z) axes are (**i**, **j**, **k**) respectively. These unit vectors at time $t+\Delta t$ (in frame $\boldsymbol{S}'$) become (**i**', **j**', **k**'). We shall consider three special rotations: the first about the x-axis, the second about the y-axis and the third about the z-axis. Referring to Fig. E.1, we suppose that in the time interval $[t, t+\Delta t]$ the frame $\boldsymbol{S}$ rotates an infinitesimal rotation $\Delta\phi$ about the x-axis. At the end of the rotation, **j** and **k** move to **j**' and **k**', respectively, however **i** remains in place because the rotation is about the x-axis. The new vectors (**i**', **j**', **k**') are described in terms of (**i**, **j**, **k**) as follows:

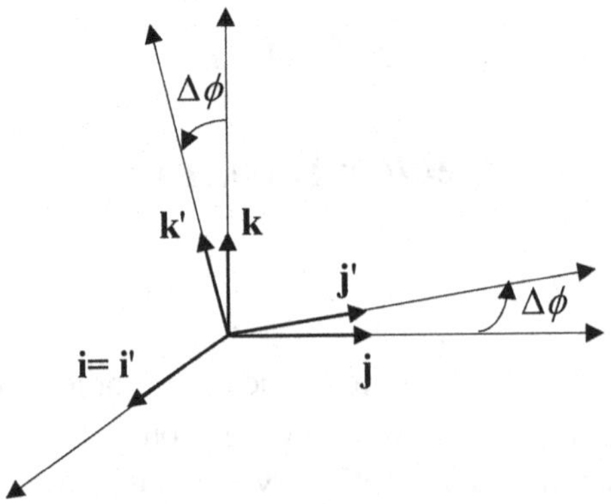

Figure E.1 Rotation of the frame about the x-axis

$$\begin{aligned}
&\mathbf{i}' = \mathbf{i}, \\
&\mathbf{j}' = \cos(\Delta\phi)\mathbf{j} + \sin(\Delta\phi)\mathbf{k} = \mathbf{j} + \Delta\phi\mathbf{k}, \\
&\mathbf{k}' = -\sin(\Delta\phi)\mathbf{j} + \cos(\Delta\phi)\mathbf{k} = -\Delta\phi\mathbf{j} + \mathbf{k}
\end{aligned} \tag{E.1}$$

Considering the kinematics of these vectors, we shall write

$$\begin{aligned}
&\mathbf{i}' = \mathbf{i}(t + \Delta t), \\
&\mathbf{j}' = \mathbf{j}(t + \Delta t), \\
&\mathbf{k}' = \mathbf{k}(t + \Delta t)
\end{aligned} \tag{E.2}$$

and from Eqs. (E.1) and (E.2) we get

$$\begin{aligned}
&\mathbf{i}(t + \Delta t)\text{-}\mathbf{i}(t) = \mathbf{0}, \\
&\mathbf{j}(t + \Delta t)\text{-}\mathbf{j}(t) = \Delta\phi\,\mathbf{k}, \\
&\mathbf{k}(t + \Delta t)\text{-}\mathbf{k}(t) = -\Delta\phi\,\mathbf{j}
\end{aligned} \tag{E.3}$$

Dividing each side by Δt and taking the limits as Δt goes to zero gives

$$\frac{d}{dt}\mathbf{i}(t) = \mathbf{0},$$
$$\frac{d}{dt}\mathbf{j}(t) = \omega_x \mathbf{k}, \qquad \text{(E.4)}$$
$$\frac{d}{dt}\mathbf{k}(t) = -\omega_x \mathbf{j}$$

where

$$\omega_x = \frac{d\phi}{dt} \qquad \text{(E.5)}$$

The above equations describe the time derivatives of the unit vectors (**i**, **j**, **k**) due to a rotation about the x-axis. Equivalently, they also describe the velocity of the tip points of these unit vectors. Repeating the above procedure for an infinitesimal angle rotation of $\Delta\theta$ about the y-axis would yield

$$\frac{d}{dt}\mathbf{i}(t) = -\omega_y \mathbf{k},$$
$$\frac{d}{dt}\mathbf{j}(t) = \mathbf{0}, \qquad \text{(E.6)}$$
$$\frac{d}{dt}\mathbf{k}(t) = \omega_y \mathbf{i}$$

where

$$\omega_y = \frac{d\theta}{dt} \qquad \text{(E.7)}$$

Finally if we perform an infinitesimal angle rotation $\Delta\psi$ about the z-axis, we would arrive at

$$\frac{d}{dt}\mathbf{i}(t) = \omega_z \mathbf{j},$$
$$\frac{d}{dt}\mathbf{j}(t) = -\omega_z \mathbf{i}, \tag{E.8}$$
$$\frac{d}{dt}\mathbf{k}(t) = \mathbf{0}$$

where

$$\omega_z = \frac{d\psi}{dt} \tag{E.9}$$

If the rotations about the (x, y, z) axes are all performed simultaneously, then the unit vector time derivatives would be obtained by summing Eqs. (E.4), (E.6) and (E.8) to get

$$\frac{d}{dt}\mathbf{i}(t) = \omega_z \mathbf{j} - \omega_y \mathbf{k},$$
$$\frac{d}{dt}\mathbf{j}(t) = \omega_x \mathbf{k} - \omega_z \mathbf{i}, \tag{E.10}$$
$$\frac{d}{dt}\mathbf{k}(t) = \omega_y \mathbf{i} - \omega_x \mathbf{j}$$

Equation (E.10) describes the velocities of the tips of the unit vectors (**i, j, k**) when frame ***S'*** rotates relative to frame ***S*** at the angular velocity $\boldsymbol{\omega}$ given by

$$\boldsymbol{\omega} = \omega_x \mathbf{i} + \omega_y \mathbf{j} + \omega_z \mathbf{k} \tag{E.11}$$

Let us consider a particle whose position vector is

$$\mathbf{r} = x\mathbf{i} + y\mathbf{j} + z\mathbf{k} \tag{E.12}$$

The time derivative of **r** would be given by

$$\begin{aligned}\frac{d}{dt}\mathbf{r} &= \frac{d}{dt}(x\mathbf{i}+y\mathbf{j}+z\mathbf{k}) \\ &= \left(\frac{dx}{dt}\mathbf{i}+\frac{dy}{dt}\mathbf{j}+\frac{dz}{dt}\mathbf{k}\right)+\left(x\frac{d}{dt}\mathbf{i}+y\frac{d}{dt}\mathbf{j}+z\frac{d}{dt}\mathbf{k}\right)\end{aligned} \tag{E.13}$$

It is seen that the vector time derivative is made of two components:

$$\mathbf{v}_S = \left.\frac{d}{dt}\mathbf{r}\right|_{S\text{ stationary}} = \left(\frac{dx}{dt}\mathbf{i}+\frac{dy}{dt}\mathbf{j}+\frac{dz}{dt}\mathbf{k}\right) \tag{E.14}$$

$$\mathbf{v}_r = \left.\frac{d}{dt}\mathbf{r}\right|_{x,y,z\text{ constants}} = \left(x\frac{d}{dt}\mathbf{i}+y\frac{d}{dt}\mathbf{j}+z\frac{d}{dt}\mathbf{k}\right) \tag{E.15}$$

From a physical standpoint, $\mathbf{v}_s$ in Eq. (E.14) is the velocity of the particle as monitored by an observer who stands in frame ***S*** (i.e. as if frame ***S*** is stationary). Substituting from Eq. (E.10) in Eq. (E.15) gives

$$\begin{aligned}\mathbf{v}_r &= x\left(\omega_z\mathbf{j}-\omega_y\mathbf{k}\right)+y\left(\omega_x\mathbf{k}-\omega_z\mathbf{i}\right)+z\left(\omega_y\mathbf{i}-\omega_x\mathbf{j}\right) \\ &= \left(\omega_y z-\omega_z y\right)\mathbf{i}+\left(\omega_z x-\omega_x z\right)\mathbf{j}+\left(\omega_x y-\omega_y x\right)\mathbf{k}\end{aligned}$$

It is straightforward to show that the above equation takes the form

$$\begin{aligned}\mathbf{v}_r &= \left(\omega_x\mathbf{i}+\omega_y\mathbf{j}+\omega_z\mathbf{k}\right)\times\left(x\mathbf{i}+y\mathbf{j}+z\mathbf{k}\right) \\ &= \boldsymbol{\omega}\times\mathbf{r}\end{aligned} \tag{E.16}$$

Equation (E.16) shows that $\mathbf{v}_r$ is the velocity of a point fixed in the rotating frame ***S*** marked at the tip of vector **r**. Now, from Eqs. (E.13)-(E.16) we can write

$$\frac{d}{dt}\mathbf{r} = \mathbf{v}_S + \boldsymbol{\omega} \times \mathbf{r} \tag{E.17}$$

Equation (E.17) is the fundamental relation for vector time derivative in a rotating frame and will be the basis for further derivatives. Taking the derivative of Eq. (E.17) yields the second derivative of **r** (the acceleration of the particle)

$$\begin{aligned}\frac{d^2}{dt^2}\mathbf{r} &= \frac{d}{dt}\left(\mathbf{v}_S + \boldsymbol{\omega} \times \mathbf{r}\right) \\ &= \frac{d}{dt}\mathbf{v}_S + \frac{d\boldsymbol{\omega}}{dt} \times \mathbf{r} + \boldsymbol{\omega} \times \frac{d\mathbf{r}}{dt}\end{aligned} \tag{E.18}$$

From Eq. (E.17) we see that

$$\frac{d}{dt}\mathbf{v}_S = \mathbf{a}_S + \boldsymbol{\omega} \times \mathbf{v}_S \tag{E.19}$$

where $\mathbf{a}_s$ the particle acceleration as observed on frame **S**. Finally, substituting from Eqs. (E.17) and (E.19) into Eq. (E.18) gives

$$\begin{aligned}\frac{d^2}{dt^2}\mathbf{r} &= \mathbf{a}_S + \boldsymbol{\omega} \times \mathbf{v}_S + \dot{\boldsymbol{\omega}} \times \mathbf{r} + \boldsymbol{\omega} \times \mathbf{v}_S + \boldsymbol{\omega} \times \left(\boldsymbol{\omega} \times \mathbf{r}\right) \\ &= \mathbf{a}_S + \dot{\boldsymbol{\omega}} \times \mathbf{r} + 2\boldsymbol{\omega} \times \mathbf{v}_S + \boldsymbol{\omega} \times \left(\boldsymbol{\omega} \times \mathbf{r}\right)\end{aligned} \tag{E.20}$$

References

1. T.C. Bradbury, Theoretical Mechanics, John Wiley & Sons, New York, 1968.
2. G. R. Fowles, Saunders College Publishing, New York, 1986.

Appendix F

Derivation of Air Speed Equations

Nomenclature:

p	air pressure
ρ	air density
v	air speed
γ	specific heat ratio
T_s	static air temperature (free stream)
T_t	total air temperature (at zero air velocity)
R	ideal gas constant
a	speed of sound
M	Mach number

The energy conservation equation, given by Bernoulli's theorem is [1-2]

$$\frac{dp}{\rho} + v dv = 0 \tag{F.1}$$

Assuming that air expands adiabatically implies

$$p = c\rho^{\gamma} \Rightarrow \rho = \left(\frac{p}{c}\right)^{\frac{1}{\gamma}} \tag{F.2}$$

Eliminating ρ from (F.1) and (F.2) gives

$$\left(\frac{c}{p}\right)^{\frac{1}{\gamma}} dp + vdv = 0 \tag{F.3}$$

Integrating (F.3) yields

$$\frac{c^{\frac{1}{\gamma}}}{1-\frac{1}{\gamma}} p^{1-\frac{1}{\gamma}} + \frac{v^2}{2} = k$$

Substituting for c from (F.2) in the above gives

$$\frac{\gamma}{\gamma-1}\frac{p}{\rho} + \frac{v^2}{2} = k \tag{F.4}$$

Using the ideal gas law $p = \rho RT_s$ in Eq. (F.4) yields

$$\frac{\gamma}{\gamma-1} RT_s + \frac{v^2}{2} = k$$

If we denote by T_t the temperature at which the air speed $v=0$, then the above equation becomes

$$\frac{\gamma}{\gamma-1} RT_s + \frac{v^2}{2} = \frac{\gamma}{\gamma-1} RT_t \Rightarrow \frac{T_t}{T_s} = 1 + \frac{(\gamma-1)}{2\gamma}\frac{v^2}{RT_s} \tag{F.5}$$

From sound wave theory

$$a^2 = \gamma RT_s \tag{F.6}$$

Since

$$M^2 = \frac{v^2}{a^2} = \frac{v^2}{\gamma R T_s} \tag{F.7}$$

Then (F.5) becomes

$$\frac{T_t}{T_s} = 1 + \frac{(\gamma - 1)}{2} M^2 \Rightarrow T_s = \frac{T_t}{1 + \frac{(\gamma - 1)}{2} M^2} \tag{F.8}$$

From the ideal gas law and Eq. (F.2) we get

$$\frac{T_t}{T_s} = \frac{p_t}{p_s}\frac{\rho_s}{\rho_t} = \frac{p_t}{p_s}\left(\frac{p_s}{p_t}\right)^{\frac{1}{\gamma}} = \left(\frac{p_t}{p_s}\right)^{1-\frac{1}{\gamma}} = \left(\frac{p_t}{p_s}\right)^{\frac{\gamma-1}{\gamma}} \tag{F.9}$$

Substituting for the temperature ratio in (F.8) gives

$$\left(\frac{p_t}{p_s}\right)^{\frac{\gamma-1}{\gamma}} = 1 + \frac{(\gamma - 1)}{2} M^2 \Rightarrow M^2 = \frac{2}{(\gamma - 1)}\left[\left(\frac{p_t}{p_s}\right)^{\frac{\gamma-1}{\gamma}} - 1\right] \tag{F.10}$$

References

1. John D. Anderson, Fundamentals of Aerodynamics, McGraw Hills, NY, New York, 1991
2. L.M. Milne-Thompson, Theoretical Aerodynamics, Macmillan & Co LTD, 1958

Appendix G

DCM Error Algebra

Even though a DCM is a 3×3 array with nine elements, it is only characterized by three parameters. The reason is that a DCM is a rotation matrix that indicates a unit vector – an axis of rotation along which a frame rotates – and an angle of rotation around this axis. Our objective is to characterize errors in the DCM in terms of these three fundamental parameters. This will be more efficient to process three elements rather than nine elements.

Errors in DCM could be classified as static or dynamic. Static sources can be attributed to initialization. Alternatively, dynamic errors result from temporal errors in the rotation vector that drives the DCM. We shall derive the equations that govern the behavior of the three parameters in both cases

Suppose $\mathbf{C}_b^a$ is the ideal DCM that transforms frame $\boldsymbol{b}$ to $\boldsymbol{a}$. The actual DCM will transform $\boldsymbol{b}$ to frame $\bar{\boldsymbol{a}}$ that is a bit different from the intended frame $\boldsymbol{a}$ and is denoted $\mathbf{C}_b^{\bar{a}}$. Supposing that the ideal and actual DCM's are related by a multiplicative matrix, **E**, i.e.

$$\mathbf{C}_b^{\bar{a}} = \mathbf{E}\mathbf{C}_b^a \tag{G.1}$$

and the error difference between the two DCM's is given by

$$\delta\mathbf{C}_b^a = \mathbf{C}_b^{\bar{a}} - \mathbf{C}_b^a = (\mathbf{E} - \mathbf{I})\mathbf{C}_b^a \tag{G.2}$$

From Eq. (G.1), **E** is given by

$$\mathbf{E} = \mathbf{C}_b^{\bar{a}} \mathbf{C}_a^b = \mathbf{C}_a^{\bar{a}} \tag{G.3}$$

Since the product of two DCM's is another DCM, then **E** can be represented by a product of three independent rotations $(\delta\alpha, \delta\beta, \delta\gamma)$. These angular rotations are assumed to be so small that small angle approximation are applicable and higher order terms are ignored, then

$$\begin{aligned}
\mathbf{E} &= \mathbf{C}_z(\delta\gamma) C_y(\delta\beta) C_x(\delta\alpha) \\
&= \begin{bmatrix} 1 & \delta\gamma & 0 \\ -\delta\gamma & 1 & 0 \\ 0 & 0 & 1 \end{bmatrix} \begin{bmatrix} 1 & 0 & -\delta\beta \\ 0 & 1 & 0 \\ \delta\beta & 0 & 1 \end{bmatrix} \begin{bmatrix} 1 & 0 & 0 \\ 0 & 1 & \delta\alpha \\ 0 & -\delta\alpha & 1 \end{bmatrix} \\
&= \begin{bmatrix} 1 & \delta\gamma & -\delta\beta \\ -\delta\gamma & 1 & \delta\alpha \\ \delta\beta & -\delta\alpha & 1 \end{bmatrix} \\
&= \mathbf{I} - \delta\tilde{\boldsymbol{\theta}}
\end{aligned} \tag{G.4}$$

where

$$\boldsymbol{\theta} = \begin{bmatrix} \alpha \\ \beta \\ \gamma \end{bmatrix} \tag{G.5}$$

Substituting for **E** from Eq. (G.4) into (G.2) yields

$$\delta \mathbf{C}_b^a = -(\delta\tilde{\boldsymbol{\theta}}) \mathbf{C}_b^a \tag{G.6}$$

Equation (G.6) shows that DCM error can be expressed as in terms of three error terms.

G.1 Dynamics of the Error Vector

Equally important is to show how the error vector propagates in time. Equation (3.34) described the time evolution of the DCM if its rotation vector is changing with time. The equation rewritten for convenience is

$$\dot{\mathbf{C}}_b^a = \mathbf{C}_b^a \tilde{\boldsymbol{\omega}}_{ab}^b$$

Applying Lemma 1.2 to the above equation gives

$$\dot{\mathbf{C}}_b^a = \mathbf{C}_b^a \left(\mathbf{C}_a^b \tilde{\boldsymbol{\omega}}_{ab}^a \mathbf{C}_b^a \right) = \tilde{\boldsymbol{\omega}}_{ab}^a \mathbf{C}_b^a = -\tilde{\boldsymbol{\omega}}_{ba}^a \mathbf{C}_b^a \tag{G.7}$$

As errors in $\boldsymbol{\omega}$ will cause errors in the DCM to accumulate, we like to describe the error propagation in terms of the error vector. Applying variation to Eq. (G.7) gives

$$\delta\dot{\mathbf{C}}_b^a = -(\delta\tilde{\boldsymbol{\omega}}_{ba}^a)\mathbf{C}_b^a - \tilde{\boldsymbol{\omega}}_{ba}^a \delta\mathbf{C}_b^a$$

Substituting for the DCM error from Eq. (G.6) in the above gives

$$\begin{aligned} \delta\dot{\mathbf{C}}_b^a &= -(\delta\tilde{\boldsymbol{\omega}}_{ba}^a)\mathbf{C}_b^a + \tilde{\boldsymbol{\omega}}_{ba}^a (\delta\tilde{\boldsymbol{\theta}})\mathbf{C}_b^a \\ &= (-\delta\tilde{\boldsymbol{\omega}}_{ba}^a + \tilde{\boldsymbol{\omega}}_{ba}^a \delta\tilde{\boldsymbol{\theta}})\mathbf{C}_b^a \end{aligned} \tag{G.8}$$

Taking the time derivative of Eq. (G.6) yields

$$\delta\dot{\mathbf{C}}_b^a = -(\delta\dot{\tilde{\boldsymbol{\theta}}})\mathbf{C}_b^a - (\delta\tilde{\boldsymbol{\theta}})\dot{\mathbf{C}}_b^a$$

Substituting in the above from Eq. (G.7) gives

$$\begin{aligned}\delta\dot{\mathbf{C}}_b^a &= -\delta\tilde{\dot{\boldsymbol{\theta}}}\mathbf{C}_b^a + (\delta\tilde{\boldsymbol{\theta}})\tilde{\boldsymbol{\omega}}_{ba}^a\mathbf{C}_b^a \\ &= \left(-\delta\tilde{\dot{\boldsymbol{\theta}}} + (\delta\tilde{\boldsymbol{\theta}})\tilde{\boldsymbol{\omega}}_{ba}^a\right)\mathbf{C}_b^a\end{aligned} \tag{G.9}$$

Equating the RHSs of Eqs. (G.8)-(G.9) and simplifying yields

$$\delta\tilde{\dot{\boldsymbol{\theta}}} = (\delta\tilde{\boldsymbol{\theta}})\tilde{\boldsymbol{\omega}}_{ba}^a - \tilde{\boldsymbol{\omega}}_{ba}^a\delta\tilde{\boldsymbol{\theta}} + \delta\tilde{\boldsymbol{\omega}}_{ba}^a$$

From Lemma 1.1, the above equation can be written as

$$\mathrm{S}(\delta\dot{\boldsymbol{\theta}}) = \mathrm{S}\left(\delta\boldsymbol{\theta}\times\boldsymbol{\omega}_{ba}^a\right) + \mathrm{S}\left(\delta\boldsymbol{\omega}_{ba}^a\right)$$

which implies that

$$\delta\dot{\boldsymbol{\theta}} = \delta\boldsymbol{\theta}\times\boldsymbol{\omega}_{ba}^a + \delta\boldsymbol{\omega}_{ba}^a \tag{G.10}$$

When needed, Eq. (G.10) can be expressed in the matrix form

$$\delta\dot{\boldsymbol{\theta}} = -\tilde{\boldsymbol{\omega}}_{ba}^a\delta\boldsymbol{\theta} + \delta\boldsymbol{\omega}_{ba}^a \tag{G.11}$$

Thus Eqs. (G.10) and (G.11) are the desired equations for describing the DCM error propagation in terms of the three-element error vector.

Appendix H

Kalman Filtering

H.1 Introduction

We might think of a stochastic process as a physical phenomenon that involves two distinctive domains: the process domain and the observation domain. The process domain provides a model that describes in some way how the process will vary with time. For example, we may be able to fashion a time periodic function that describes the daily temperature at a certain geographical spot. Of course, this will not mean that the temperature at any time will be as prescribed by the model or even be closer to it; but this might provide a guideline to what the temperature will be. Therefore we strive to find a mathematical model that shows how the process varies with time.

The stochastic process will evolve with or without our knowledge (provided that we do not interfere with it), but we will be completely oblivious to it unless we have some means of observing the phenomenon. Continuing with our example, we will be able to know how the temperature varies only if we have a thermometer at this spot of interest. We might include a pressure and humidity sensors to see what kind of influence, if any, on the behavior of the temperature variation.

With the process domain, there are uncertainties that prevent the process from following the mathematical model. Likewise, no matter how accurate the thermometer is, there will be uncertainties about its measurements. Before we continue, we elaborate a bit on the concept of the 'state'. The state of a stochastic process is the smallest set of relevant variables that, if known, will describe the stochastic process completely. For example, if the pressure and temperature of some gas are members of the state, then we do not need to include the density. Likewise, in a linear electric circuit including the voltage in the state obviate the need for including the current.

H.2 Linear Kalman Filter

In developing the Kalman filter [1-4], we will deal with a specific type of stochastic processes for which the state temporal behavior is characterized by the discrete-time Gauss-Markov process governed by

$$\mathbf{x}_{i+1} = \mathbf{F}_i \mathbf{x}_i + \mathbf{G}_i \boldsymbol{\eta}_i, \quad i = 1, 2, \cdots, \tag{H.1}$$

In the above equation, **x** is the state vector whose dimension is determined by the number of the variables that describe the process, **F** is the state transition matrix, and **G** is the input coefficients matrix. The process is driven by the zero mean Gaussian white noise $\boldsymbol{\eta}$. The index i, denotes the time epoch at which a variable is given. When this index starts at 1, it only signifies that we started observing the process at this point in time, but the process itself could be running much earlier.

The statistics of the above process are governed by

$$\mathrm{E}\boldsymbol{\eta}_i = \mathbf{0} \tag{H.2a}$$

$$\mathrm{E}\boldsymbol{\eta}_i \boldsymbol{\eta}'_j = \mathbf{Q}_i \delta_{ij} \tag{H.2b}$$

$$\mathrm{E}\mathbf{x}_i \boldsymbol{\eta}'_j = \mathbf{0}, \quad \forall i,\ j \le i \tag{H.2c}$$

where the operator E denotes the statistical expected value and δ is the kronecker delta. At 'time' 0 it shall be assumed that the mean value of the state (given all the previous measurements)

$$\bar{\mathbf{x}}_0 = \mathrm{E}\mathbf{x}_0 \tag{H.2d}$$

and its covariance matrix

$$\bar{\mathbf{P}}_0 = \mathrm{E}(\mathbf{x}_0 - \bar{\mathbf{x}}_0)(\mathbf{x}_0 - \bar{\mathbf{x}}_0)' \tag{H.2e}$$

are known as they will be needed to initialize the state estimate and its covariance. It will be assumed the state $\mathbf{x}$ is observed by a set of measurements given by the vector $\mathbf{y}$ which is linear of $\mathbf{x}$ through the matrix $\mathbf{H}$. The dimensions of $\mathbf{x}$ and $\mathbf{y}$ in general are different and not restricted to any size. We also assume that $\mathbf{y}$ is corrupted with a Gaussian white noise vector $\mathbf{v}$ of the same dimension as $\mathbf{y}$. Thus the measurement equation is

$$\mathbf{y}_i = \mathbf{H}_i\mathbf{x}_i + \mathbf{v}_i, \quad i = 1, 2, \cdots, \tag{H.3}$$

In the above equations the statistics of the process $\mathbf{v}$ are governed by

$$\mathrm{E}\,\mathbf{v}_i = \mathbf{0} \tag{H.4a}$$

$$\mathrm{E}\,\mathbf{v}_i\mathbf{v}'_j = \mathbf{R}_i\delta_{ij} \tag{H.4b}$$

$$\mathrm{E}\,\mathbf{x}_i\mathbf{v}'_i = \mathbf{0} \tag{H.4c}$$

The dynamic model of the process and the measurement is depicted in Figure H.1.

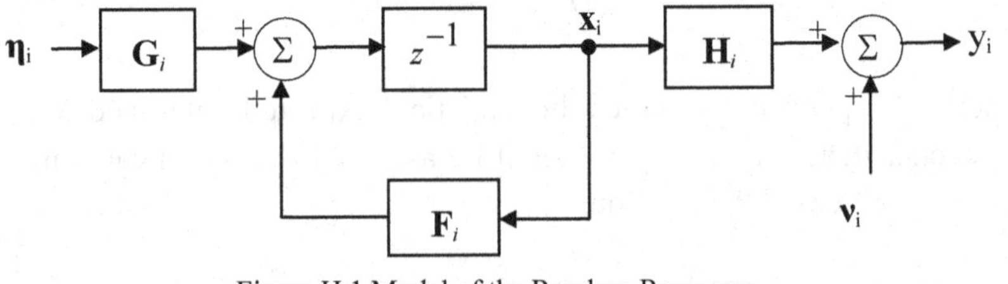

Figure H.1 Model of the Random Processes

The objective of the linear estimator is to estimate $\mathbf{x}_n$ at time n using linear combination of the collected n measurements $\mathbf{y}_1, \mathbf{y}_2, \ldots, \mathbf{y}_n$. That is

$$\hat{\mathbf{x}}_n = \mathbf{A}_1\mathbf{y}_1 + \mathbf{A}_2\mathbf{y}_2 + \cdots + \mathbf{A}_n\mathbf{y}_n \tag{H.5}$$

The coefficients $\mathbf{A}_1$, $\mathbf{A}_2,\ldots,\mathbf{A}_n$ are selected so that $\hat{\mathbf{x}}_n$ is a minimum variance estimate, i.e. $\mathrm{E}(\mathbf{x}_n - \hat{\mathbf{x}}_n)'(\mathbf{x}_n - \hat{\mathbf{x}}_n)$ is a minimum. The estimator equations are implemented in a rather elegant recursive algorithm given by the set of equations given below.

H.2.1 The Gain Equation

The filter is mathematically structured so that corrections to the state (when a new measurement is received) are linear to this measurement through the gain matrix

$$\mathbf{K}_n = \overline{\mathbf{P}}_n\mathbf{H}'_n\left(\mathbf{R}_n + \mathbf{H}_n\overline{\mathbf{P}}_n\mathbf{H}'_n\right)^{-1} \tag{H.6}$$

The use of the gain and applying the corrections are shown in the following.

H.2.2 The Measurements Update Equations

At epoch, n, the filter receives a new measurement, $\mathbf{y}_n$, and uses it to correct the state from $\overline{\mathbf{x}}_n$ to $\hat{\mathbf{x}}_n$ and it computes the covariance $\hat{\mathbf{P}}_n$ of the new filtered state, as follows:

$$\hat{\mathbf{x}}_n = \overline{\mathbf{x}}_n + \mathbf{K}_n\left(\mathbf{y}_n - \mathbf{H}_n\overline{\mathbf{x}}_n\right) \tag{H.7}$$

$$\begin{aligned}\hat{\mathbf{P}}_n &= \mathrm{E}(\mathbf{x}_n - \overline{\mathbf{x}}_n)(\mathbf{x}_n - \overline{\mathbf{x}}_n)' \\ &= \overline{\mathbf{P}}_n - \overline{\mathbf{P}}_n\mathbf{H}'_n\left(\mathbf{R}_n + \mathbf{H}_n\overline{\mathbf{P}}_n\mathbf{H}'_n\right)^{-1}\mathbf{H}_n\overline{\mathbf{P}}_n\end{aligned} \tag{H.8}$$

H.2.3 The Time Propagate Equations

At the present epoch, n, the filter 'predicts' what the state would be at the next epoch, n+1, and computes its covariance. The predicted state $\overline{\mathbf{x}}_{n+1}$ and its covariance $\overline{\mathbf{P}}_{n+1}$ are computed by the equations

$$\overline{\mathbf{x}}_{n+1} = \mathbf{F}_n \hat{\mathbf{x}}_n \tag{H.9}$$

and

$$\begin{aligned} \overline{\mathbf{P}}_{n+1} &= \mathrm{E}(\mathbf{x}_{n+1} - \hat{\mathbf{x}}_{n+1})(\mathbf{x}_{n+1} - \hat{\mathbf{x}}_{n+1})' \\ &= \mathbf{F}_n \hat{\mathbf{P}}_n \mathbf{F}'_n + \mathbf{G}_n \mathbf{Q}_n \mathbf{G}'_n \end{aligned} \tag{H.10}$$

The flow diagram of the above equations is depicted in Fig. H.2.

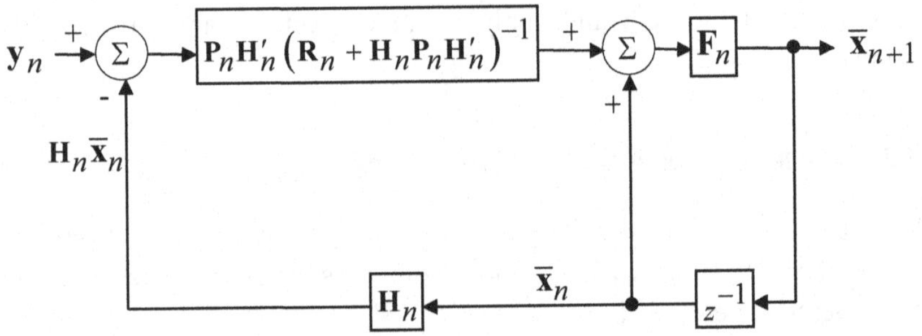

Figure H.2 Linear Estimator Flow Diagram

H.3 Non Linear Estimation

More often than not, practical estimation problems are nonlinear. The state and measurement equations typically take the form of

$$\mathbf{x}_{n+1} = \mathbf{f}(\mathbf{x}_n) + \mathbf{u}_n \tag{H.11}$$

$$\mathbf{y}_n = \mathbf{h}(\mathbf{x}_n) + \mathbf{v}_n \tag{H.12}$$

To apply the linear estimation tools, the above equations are linearized to a form similar to those in Eqs. (H.1) and (H.3) [2]. This is discussed in the following.

H.3.1 Linearization

Suppose we have a nominal trajectory, $\mathbf{x}_{nom}$, for which

$$\mathbf{x}_{nom,n+1} = \mathbf{f}(\mathbf{x}_{nom,n}) \tag{H.13}$$

$$\mathbf{y}_{nom,n} = \mathbf{h}(\mathbf{x}_{nom,n}) \tag{H.14}$$

Differencing Eq. (H.11) from (H.13) and Eq. (H.12) from (H.14) gives

$$\mathbf{x}_{n+1} - \mathbf{x}_{nom,n+1} = \mathbf{f}(\mathbf{x}_{nom,n}) - \mathbf{f}(\mathbf{x}_n) + \mathbf{u}_n \tag{H.15}$$

$$\mathbf{y}_n - \mathbf{y}_{nom,n} = \mathbf{h}(\mathbf{x}_{nom,n}) - \mathbf{h}(\mathbf{x}_n) + \mathbf{v}_n \tag{H.16}$$

Let

$$\begin{aligned} \delta\mathbf{x}_n &= \mathbf{x}_n - \mathbf{x}_{nom,n} \\ \delta\mathbf{y}_n &= \mathbf{y}_n - \mathbf{y}_{nom,n} \end{aligned} \tag{H.17}$$

Linearizing the functions **f** and **h** about the nominal trajectory and applying to Eqs. (H.15) and (H.16) gives

$$\delta\mathbf{x}_{n+1} = \mathbf{F}(\mathbf{x}_{nom,n})\delta\mathbf{x}_n + \mathbf{u}_n \tag{H.18}$$

$$\delta \mathbf{y}_n = \mathbf{H}(\mathbf{x}_{nom,n})\delta \mathbf{x}_n + \mathbf{v}_n \tag{H.19}$$

where

$$\mathbf{F}_{ij}(\mathbf{x}_{nom,n}) = \left.\frac{\partial \mathbf{f}_i(\mathbf{x})}{\partial \mathbf{x}_j}\right|_{\mathbf{x} = \mathbf{x}_{nom,n}} \tag{H.20}$$

$$\mathbf{H}_{ij}(\mathbf{x}_{nom,n}) = \left.\frac{\partial \mathbf{h}_i(\mathbf{x})}{\partial \mathbf{x}_j}\right|_{\mathbf{x} = \mathbf{x}_{nom,n}} \tag{H.21}$$

In the above two equations the subscript ij denotes the ith row and the jth column.

Remarks: Implied in the above formulation that

1. **f** and **h** are continuous and differentiable functions in **x**.
2. We are estimating the error in the state (relative to the nominal trajectory) rather than the state itself.
3. The noise terms **u** and **v** are Gaussian white noise.
4. Higher order linearization terms are ignored in the above two equations.
5. There are two ways for implementing the estimator equations for nonlinear system: the open loop and the closed loop methods. Both are addressed below.

Using a nominal trajectory as shown in Eqs. (H.18) and (H.19) would require storing all the data for the intended trajectory. This could impose severe demands on the computer resources. Alternatively, one could use the state estimate in place of the nominal trajectory. This implies that the linearized matrix coefficients would be

$$\mathbf{F}_{ij}(\mathbf{x}_{nom,n}) = \left.\frac{\partial \mathbf{f}_i(\mathbf{x})}{\partial \mathbf{x}_j}\right|_{\mathbf{x}=\hat{\mathbf{x}}_n} \tag{H.22}$$

$$\mathbf{H}_{ij}(\mathbf{x}_{nom,n}) = \left.\frac{\partial \mathbf{h}_i(\mathbf{x})}{\partial \mathbf{x}_j}\right|_{\mathbf{x}=\hat{\mathbf{x}}_n} \tag{H.23}$$

With these linearized coefficients, we can implement the estimator equations in two ways, the open loop (feedforward) and the closed loop (feedback) implementations [3].

H.3.2 Open Loop Implementation

In the open loop implementation, we estimate and propagate the error in the state. The error may be added to the state when desired to have an estimate of the state itself. The state vector is updated as if it is a nominal trajectory that is not influenced by the state estimation. Implementation of this open loop method is depicted in Fig. H.3.

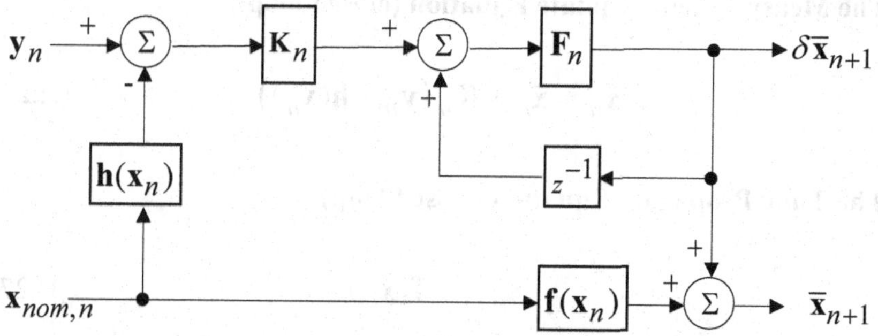

Figure H.3 Open Loop Non Linear Estimator Flow Diagram

The estimator state equation updates Eqs. (H.7) and (H.9) now become:

The Measurements Update Equation (open loop):

$$\delta\hat{\mathbf{x}}_n = \delta\bar{\mathbf{x}}_n + \mathbf{K}_n\left(\mathbf{y}_n - \mathbf{h}(\mathbf{x}_{nom,n})\right) \tag{H.24}$$

The Time Propagate Equation (open loop):

$$\delta\bar{\mathbf{x}}_{n+1} = \mathbf{F}_n\delta\hat{\mathbf{x}}_n \tag{H.25}$$

The remainder of the state estimator equations (H.6), (H.8) and (H.10) will be the same. As shown in Fig. H.3 the state and measurement equations of the nominal trajectory are updated by Eqs. (H.13) and (H.14).

H.3.3 Closed Loop Implementation

In the closed loop method, the measurement update is added to the state. Therefore the state is continually updated with every new measurement. The flow diagram is depicted in Fig. H.4. The estimator state equation updates in Eqs. (H.7) and (H.9) now become

The Measurements Update Equation (closed loop):

$$\hat{\mathbf{x}}_n = \bar{\mathbf{x}}_n + \mathbf{K}_n\left(\mathbf{y}_n - \mathbf{h}(\bar{\mathbf{x}}_n)\right) \tag{H.26}$$

The Time Propagate Equation (closed loop):

$$\bar{\mathbf{x}}_{n+1} = \mathbf{f}(\hat{\mathbf{x}}_n) \tag{H.27}$$

Like in the open loop approach, the state estimator equations (H.6), (H.8) and (H.10) will be the same.

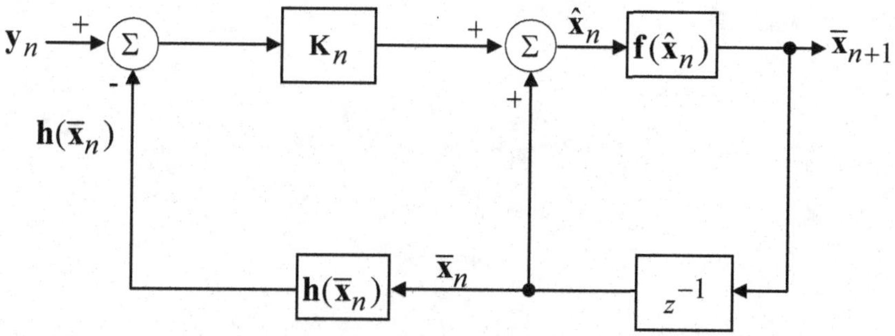

Figure H.4 Closed Loop Non Linear Estimator Flow Diagram

References

1. R. E. Kalman, "A New Approach to Linear Filtering and Prediction Problems," J. Basic Eng., A.S.M.E., 82, 1960, pp. 35. 45.
2. R. S. Bucy and P. D. Joseph, Filtering for Stochastic Processes with Applications to Guidance, John Wiley, Interscience Publishers, N. Y., New York, 1968.
3. J. A. Farrell & M. Barth, The Global Positioning System & Inertial Navigation, McGraw-Hill, New York, 1999.
4. C. Jekeli, Inertial Navigation Systems with Geodetic Applications, Walter de Gruyter, Berlin, 2001.

Index

acceleration bias, 149
AHRS (Attitude Heading and Reference System), 152
air speed, 104, 105, 115, 117, 118
alignment, 136, 137, 151
altitude rate, 117
attitude equation, 78

clock bias, 188, 190
craft angular rates, 68
craft rate, 169
craft velocity, 86, 95

declination angle (magnetic), 155, 159
density altitude, 116
direction cosine matrix, 19
    differential equation, 48

Earth
    eccentricity, 62, 63
    flattening, 62, 63
    magnetic field, 155, 156
    radius of curvature, 64
    rate, 67, 170, 171
ECEF, 180, 183, 185, 188, 190, 191
ellipsoid, 59, 60, 62, 63, 65, 66, 68
    ellipsoid height, 60
equipotential surface, 59, 60, 71

error state, 168, 179, 180, 183, 191
Euler angles, 35, 37, 44, 45, 47, 48, 50, 55, 56
    differential equation, 55

Frame
    body frame, 76, 78, 79, 82, 83
    Earth frame, 65, 66, 68, 76, 81, 84, 85
    inertial frame, 65, 76
    magnetic frame, 156
    navigation frame, 65, 66, 76, 81, 82
frame rotations, 36

g slaving, 160
geoid, 60, 73
    geoid elevation, 60
geopotential height, 106, 107, 111
Global Positioning System (GPS), 167, 168, 187, 188, 190, 191, 192
gravity, 58, 59, 60, 63, 72, 73
gyro bias compensation, 160
gyro drift, 136, 149, 159

horizontal intensity(magnetic), 155
hydrostatic equation, 108

inclination angle(magnetic), 155
indicated air speed, 115, 119
inertial navigation error equation, 185
Inertial Navigation System (INS), 167, 168, 190, 191
inertial sensor
    accelerometer, 76, 77, 82, 83
    rate gyro, 77, 78, 83
inner transformation matrix, 29

Kalman filter, 167, 168, 185, 190, 191, 192

lapse rate, 105, 109, 110, 116
latitude
    geocentric latitude, 63
    geodetic latitude, 63, 65, 66, 71

Mach number, 115, 117, 118
maneuver detector, 165
matrix, 7, 8, 10, 11, 12, 14, 15, 16
    column, 8
    fixed axis, 24, 25, 28
    identity, 10
    orthonormal, 9, 10, 14, 15
    row, 8
    skew symmetric matrix, 11
    transpose, 8, 10
    unitary, 20, 24, 28
mean sea level, 59, 60
meridian, 62, 63, 64, 66

navigation equation, 80, 82, 83, 84, 85
normal ellipsoid, 60, 71

outer transformation matrix, 29

polar circle, 126, 128, 130, 132, 133, 134
pressure altitude, 104, 105, 110, 111, 113, 114, 117

quaternion, 33, 35, 38, 39, 40, 41, 42, 43, 44, 45, 46, 47, 48, 52, 55, 56
    differential equation, 50
    identity, 39
    inverse, 39
    norm, 39, 40
    normalized, 40

Rodrigues formula, 28
rotation matrix, 26, 28
rotation vector, 28, 33,35, 38, 39, 42, 45, 46, 50, 52, 54, 55, 56
    differential equation, 52, 57, 87, 91

Somigliana formula, 71
Standard Atmosphere (US) 1976, 105, 119

total intensity(magnetic), 155

transformation matrix, 21, 25, 26, 28, 29
true altitude, 105, 106, 107, 111, 113, 114

vector
    cosine law, 9
    cross product, 11, 12, 16
    Hermitian inner product, 10
    inner product, 7, 9, 10
    length, 9
    norm, 9
    orthogonal, 7, 9, 11
    orthonormal, 9, 10, 14, 15
vertical channel, 101, 103

wander azimuth, 123, 124, 125, 126, 127, 132
World Geodetic System, 62, 74
World Magnetic Model, 155, 166